Hugh Fraser

Handy Book of Ornamental Conifers of Rhododendrons and other AmericanFlowering Shrubs Suitable for the Climate and Soils of Britain

Hugh Fraser

Handy Book of Ornamental Conifers of Rhododendrons and other AmericanFlowering Shrubs Suitable for the Climate and Soils of Britain

ISBN/EAN: 9783337225889

Printed in Europe, USA, Canada, Australia, Japan

Cover: Foto ©berggeist007 / pixelio.de

More available books at **www.hansebooks.com**

HANDY BOOK

OF

ORNAMENTAL CONIFERS

ETC.

PRINTED BY WILLIAM BLACKWOOD AND SONS, EDINBURGH.

HANDY BOOK

OF

ORNAMENTAL CONIFERS

AND OF

RHODODENDRONS AND OTHER AMERICAN FLOWERING SHRUBS

SUITABLE FOR THE CLIMATE AND SOILS
OF BRITAIN

BY

HUGH FRASER

FELLOW OF THE BOTANICAL SOCIETY OF EDINBURGH

WILLIAM BLACKWOOD AND SONS
EDINBURGH AND LONDON
MDCCCLXXV

TO

THOMAS METHVEN, ESQ.

OF

THE LEITH WALK NURSERIES, EDINBURGH

THIS VOLUME

IS MOST RESPECTFULLY DEDICATED

BY

THE AUTHOR

PREFACE.

There is perhaps no feature of the present age more strikingly obvious, or more hopeful as an evidence of the growing taste and culture of the people, than the almost universal interest now taken in matters connected with Botany and Horticulture.

The number of amateur cultivators of plants and flowers has greatly increased during the last few years, and the very success which they have attained has created a desire for further knowledge, and a demand for that class of literature so admirably represented by such works as the 'Handy Book of the Flower-Garden,' 'Herbaceous and Alpine Plants,' and various others, all of which give the result of the long experience and keen observation of their authors in a form at once comprehensive, practical, and popular.

So far as the Author is aware, no book written on a similar plan, dealing with the beautiful trees and shrubs which form the subject of this volume, has hitherto been published; he has been constrained, therefore, to attempt to supply this acknowledged desideratum, and he believes that the young gardener as well as the amateur will find in his "Handy Book" numerous facts and suggestions calculated to assist them in the successful cultivation of the various species and varieties therein described.

EDINBURGH, *September 1875.*

CONTENTS.

ORNAMENTAL CONIFERS.

	PAGE		PAGE
Introduction,	1	Podocarpus,	122
Abies,	9	Prumnopytis,	125
Araucaria,	25	Retinospora,	126
Biota,	29	Salisburia,	134
Cedrus,	35	Saxe-Gothæa,	136
Cryptomeria,	41	Sciadopytis,	137
Cephalotaxus,	43	Sequoia,	138
Chamæcyparis,	47	Taxus,	140
Cupressus,	49	Taxodium,	148
Fitz-Roya,	59	Thuja,	150
Juniperus,	60	Thujopsis,	156
Larix,	72	Torreya,	159
Libocedrus,	78	Wellingtonia,	161
Pinus,	81	Widdringtonia,	167
Picea,	107		

RHODODENDRONS AND OTHER AMERICAN OR PEAT-SOIL SHRUBS, ... 168

Rhododendrons,	170	Bryanthus,	225
Azalea,	210	Clethra,	226
Andromeda,	217	Comptonia,	227
Arctostaphylos,	223	Chamælodon,	229

Daphne, . . .	230	Ledum, . . . 253
Dirca, . . .	234	Menziesia, . . . 257
Erica, . .	235	Myrica, . . . 261
Empetrum, .	243	Pernettya, . . . 263
Epigæa, .	245	Polygala, . . . 265
Gaultheria, . .	246	Rhodora, . . . 266
Kalmia, .	248	Vaccinium, . . . 268

HERBACEOUS PLANTS Suitable for Associating with Peat-Soil Shrubs, ... 273

Alstrœmeria, .	274	Helleborus, .	. 280
Arundo, .	275	Hepatica, . .	. 280
Arum, . . .	275	Leucojum, .	. 281
Asclepias, . .	276	Lilium, .	. 281
Bulbocodium, .	276	Osmunda, .	. 284
Campanula,	277	Phormium, .	285
Colchicum, . .	278	Struthiopteris,	286
Crocus, .	278	Tritoma, .	. 286
Gentiana,	278	Tigridia,	286
Gladiolus,	279	Trillium,	. 287
Gynerium, . .	279	Yucca,	287
Galanthus, .	279		

INDEX, 289

ORNAMENTAL CONIFERS.

INTRODUCTION.

AMONG the many orders into which the Vegetable Kingdom is divided, it would be difficult to name one more important, either from an economic or ornamental point of view, than the Coniferæ. Their wonderful diversity of habits—from the procumbent Junipers of Europe, whose stature is measured by inches, to the majestic *Wellingtonia* of California, rising to the almost fabulous height of 360 feet— the varieties of colour and tint which the foliage of the various species assume, and the uniform gracefulness and symmetry of growth which characterise every member of the family, combined with the fact that a large proportion of the grandest forms are adapted to the soils and climate of this country, render them universal favourites as ornamental plants; while the well-known value of many of them as timber-trees must always insure their extensive use in forest-planting.

Ornamental Conifers—in which we include all the sorts hardy enough for the open air in Britain, not even excepting those cultivated as forest-trees, seeing that they are for the most part quite as handsome as the rarest of the new introductions—are usually planted as single specimens promiscuously with the other decorative plants on lawns and pleasure-grounds, creating a richness and grandeur of effect which cannot be surpassed; or where a large collection is desired, and the necessary space can be afforded, they are now very frequently grown by themselves in a portion of the grounds specially devoted to them, called a Pinetum, where their several habits and peculiarities may be contrasted, and their characters studied, with greater facility than where they are diffused over a more extensive area. Of these there are many admirable examples throughout the country; and no portion of the pleasure-grounds is more interesting, or more appreciated both by the proprietor and visitors, than the Pinetum.

In the planting and arrangement of a collection of coniferous trees and shrubs—whether in the Pinetum or in the ordinary pleasure-grounds—so as to produce, by the blending of the various forms and shades of colour, that picturesque beauty for which they supply such abundant material, much must necessarily depend upon the taste and skill of the planter, and the variety of sorts at his disposal, as well as the extent and character of the grounds to be operated on. It is at the same time to be noticed, that there are principles of general application by which he must as far as possible be guided, if his labours are to be followed by the fullest amount of success.

Prominent among these principles is the choice of a suitable situation, which, whenever it is possible, should on the one hand be open, airy, and fully exposed to the sun; and on the other, sheltered enough to protect the plants from violent winds, which are fatal to even the hardiest of the tribe. Nothing is more common than to see fine varieties planted where they are literally blown out of shape, the sides of the plants most exposed withered and leafless, making little or no growth, and after a short and miserable existence, dying off from time to time, or soon becoming such unsightly objects as to render their removal a necessity. Nor are the evils of planting in low confined situations less serious. In such they rarely mature their wood so as to fortify them against the rigours of an average winter, and the result appears in their being more or less injured by an amount of frost which in other circumstances would not affect them in the slightest degree; and even if, owing to an exceptionally dry summer and autumn, they are enabled to fully ripen their wood and to pass scathless through the winter, the no less severe ordeal of late spring frosts awaits them, to which their tendency in such situations to start early into growth makes them peculiarly susceptible.

Another very common cause of failure in the success of coniferous plants is imperfect drainage; and in all cases where, from the land being naturally damp, or the subsoil being stiff and retentive, there is any liability to the accumulation or stagnation of water at the roots, it should be thoroughly drained before planting.

As may be inferred from the fact that Pinetums have been formed throughout the country, in which representatives of all the hardy genera are grown side by side with little or no variety of soil, and generally with great success, the Coniferæ are, on the whole, most accommodating, as far as mere soil is concerned, and a large proportion of the species will grow and thrive on almost any geological formation. It is necessary to remark, however, that some are more fastidious, and have certain preferences or requirements, and thrive in proportion to the extent to which these are either naturally or artificially supplied. To some of these we will advert in the course of our notes on the species;

but meantime it may safely be said that, with land so drained as neither to be liable to excessive moisture nor drought, a porous subsoil, whether its composition be sand, gravel, or chalk, and a rich, deep, sandy loam, will fully supply the wants of nine-tenths of the hardy species.

In common with most other plants, Conifers delight in rich, well-decomposed manure; and where, from the poverty of the soil, they have a tendency to become stunted and unhealthy, much may be done to restore their vigour by its being liberally supplied, either alone or mixed with well-rotted leaf-mould and turfy loam, forked or dug in at such a distance from the stem that the young fibres may at once avail themselves of it.

All the sorts may be successfully transplanted, either in early autumn, immediately after the wood begins to ripen, or in spring, before the growth commences. We are satisfied, however, that autumn is, after all, the best season for the purpose, and would recommend the work being begun as early in September as the state of the young wood will admit of, and to be finished by the end of November. The soil is then neither chilled by frost nor likely to be saturated by long-continued rains; the roots are still active, and have yet time to make a little progress and establish themselves in their new quarters before winter sets in. It is at the same time true, that when the plants are of a moderate size, and so prepared by frequent removals that they can be lifted with firm balls, and have abundance of small, fibry roots, the season of planting is of comparatively little importance, as they may be moved with perfect safety any time between September and April, provided the weather is fresh, and the soil sufficiently dry to be firmed about the roots without being soddened.

The process of transplanting is one that, though in itself simple and easily understood, is nevertheless so important for the wellbeing of Conifers as to demand the most careful attention on the part of the operator; and as many failures are to be attributed to its being unskilfully or carelessly performed, a few suggestions may be useful. The pits should be made much larger than the balls, so as to admit of the roots being regularly and fully spread out on every side, leaving a good margin for the introduction of soft soil, or, if the ground is poor, prepared compost, among and around them. The bottom should be well dug up, a portion of the subsoil removed, and replaced by the surface soil or compost. The plant must not be inserted deeper than to allow of the upper roots being covered with a few inches of soil; deep planting is injurious to most plants, and specially so to Conifers, and should be carefully avoided. After the roots have been properly spread out and adjusted, and the necessary fine soil or compost introduced, the pits should be filled with earth, and thereafter well firmed round the stem with the feet. If the soil is dry, it will be necessary to make a basin round each plant, and to give it a thorough watering before the surface is finally levelled and dressed. When the work is deferred till late in spring, and is followed by dry, scorching weather, it will be necessary to repeat the watering, and even to syringe them freely overhead both morning and evening, until they show signs of a movement in the way of growth. In such circumstances a slight mulching of leaves, half-rotten manure, or short grass will be found beneficial in preventing evaporation, and may be left on through the summer, or at least until such time as the weather becomes damp and wet, with advantage to the plants, as well as saving a great amount of labour.

In view of the varied heights and habits of growth which prevail among the species, it is obvious that no general rule can be laid down as a guide to the distance at which a mixed collection of coniferous trees and shrubs should be planted apart. The planter must consider the character of each individual, and the space which it may be expected to occupy when, in the course of years, it shall attain to something like maturity, and arrange accordingly.

Many mistakes have been made in regard to this; and some of what would otherwise have been the finest specimens in the country have been irretrievably damaged by being planted too closely together, or too near buildings, fences, and walks, and thus prevented from developing themselves, and assuming their natural characters. When, as is frequently the case, it is necessary to plant at first thickly, or to introduce among them other trees, either for the purpose of shelter or for clothing the ground so as to produce an immediate effect, the most careful attention should be given year after year to thinning, so that the growth of the permanent specimens may not be interfered with, and that each may have abundance of room on every side for extending its branches.

If properly planted, and under tolerably favourable conditions as regards soil and situation, Conifers are not very exacting upon the care and attention of the cultivator. He has, however, duties to perform which may not be neglected without serious injury to the plants. A Pine, for example, will occasionally be found from some accident losing its leading shoot, a contingency which must immediately be met by an endeavour to induce one of the laterals to take its place by tying it up, in which he will in most cases succeed. Two or three leaders will sometimes make their appearance, and strive for the mastery: these, with the exception of the strongest, must be stopped or removed as soon as they are observed; and if taken at an early stage, they will be effectually dealt with by a simple twist with the finger and thumb, which, by arresting their progress, will allow the selected leader to take its proper position. The pruning or stopping of such branches as have a tendency to spoil the symmetry of the plant must also be duly attended to—an operation of great importance, and one which may be freely performed on the great majority of the species, in many instances, with the greatest advantage, and for which the best season is early autumn, when, the growth being fully matured, they are least liable to bleed, and the wounds have abundance of time to heal before the rising of the sap in spring.

ABIES (THE SPRUCE FIRS).

Distinguished from the true Pines by their pendent cones and short solitary leaves, in some species scattered irregularly round the shoots, and in others more or less distinctly in rows, as well as by their peculiarly erect conical habits of growth, the Spruce Firs rank among the finest of our evergreen trees.

They are for the most part natives of the colder regions of Europe, Asia, and America, where they extend to the utmost limits of woody vegetation, forming immense forests, thriving in the soils of almost every geological formation, but attaining their greatest heights and perfection when sheltered from the blast, and where the soil is of a peaty or rich alluvial character.

Most of the known species are quite equal to the climate of Britain, doing well in almost any situation, if not too dry.

All the really hardy sorts are useful for decorative planting, and worthy of being extensively introduced into parks and pleasure-grounds, where they form handsome single specimens; while their bold outlines and distinct colours render them invaluable for producing striking landscape effects, either when grouped in masses, blended with other trees, or as single specimens in parks or lawns.

We note the following as among the most distinct species and varieties in cultivation:—

A. **Albertiana** (*Prince Albert's Spruce*), also known as *Mertensiana*, was introduced about twenty years ago from Oregon, where, as well as in Northern California, it is found occupying vast tracts of country, and rising to heights of from 100 to 150 feet, forming a handsome, bushy-headed tree, with a straight, tapering stem, from 4 to 6 feet in diameter near the ground. In general appearance it so closely resembles the Hemlock Spruce (*A. canadensis*) that it was at first regarded as a variety of that species:

it is, however, quite distinct, having a stronger constitution and more rapid growth, while the leaves are shorter, more slender, and produced in greater profusion. It is thoroughly hardy in this country, and does well wherever the soil is moderately moist and contains a good proportion of decayed vegetable matter. Few plants are better adapted for lawns, or any situation requiring single specimens—its graceful drooping branches, and neat pyramidal outline, rendering it always attractive.

A. alba (*the White Spruce*).—Indigenous to the colder parts of Canada and the United States, extending to within twenty miles of the Arctic Sea, seldom rising higher than from 30 to 50 feet, with a straight stem of from 1 to 2 feet in diameter near the ground: was introduced into Britain in the beginning of the last century. It is one of the hardiest of the species, thriving in the bleakest situations, and in most soils if not very dry, but preferring a somewhat shady situation or northern aspect. It forms a handsome, conical tree, regularly and thickly furnished with branches from the ground upwards. The leaves, which are thickly scattered over the branches, are of a very light glaucous tint, giving the plant a distinct silvery appearance.

Of several varieties, the two following are the most distinct:—

Var. *glauca* (*the Glaucous-leaved White Spruce*) has

much brighter foliage, and a habit of growth, though scarcely so robust, very much the same as the species.

Var. *nana* (*the Dwarf White Spruce*).—This, like the preceding, is of European origin, and is peculiar from its dwarf habit of growth, forming a close bush of from 2 to 3 feet in height: it is interesting as a rockery-plant, and forms a neat specimen for a small lawn or front of a shrubbery border.

A. Alcoquiana (*Alcock's Spruce*), sent home in 1860 by Mr J. G. Veitch from the Fusi-yami Mountains in Japan, where he found it at elevations of from 5000 to 7000 feet above the level of the sea, growing to heights of from 80 to above 100 feet.

Though as yet a comparative stranger, and not very widely known in British gardens, it has proved itself, as far as opportunity has been had of testing it, to be quite hardy; and being of free growth and of a neat, dense, conical habit, and moreover quite distinct in appearance from any of the other species, it is without doubt a valuable addition to our ornamental Conifers, and worthy of a trial in every collection. It thrives best in a deep, rich soil, and should have the benefit of a moderately sheltered situation.

A. canadensis (*the Hemlock Spruce*).—This tree is remarkable for the vast extent of its range in Canada and the colder parts of the United States, extending from Carolina to the borders of Hudson Bay, and from the Atlantic to the Pacific Ocean, frequently forming the principal arborescent vegetation for hundreds of miles, growing to heights of from 30 to 80 feet, and always seen in its greatest perfection on the edges of swamps or in deep, moist, alluvial soils. Although coarse-grained and not very durable, its timber is extensively used in America for temporary purposes; while the bark, in combination with that of the oak, is invaluable for tanning.

It has been cultivated in this country for upwards of 100 years, and is much valued for ornamental planting. The branches are very numerous, slender, and drooping, giving the tree a graceful, feathery appearance. The leaves are broad, set in two rows, of a light-green tint on the upper and glaucous on the under surface. It requires a moist, rich soil, and a sheltered situation, growing in such circumstances with great vigour, and forming a beautiful, bushy, broad-headed specimen tree.

A. Douglasii (*Douglas's Spruce*). — Though discovered at the end of the last century by Menzies at Nootka Sound, the honour of introducing this magnificent tree is due to David Douglas, in compliment to whom it was named, and to whose indefatigable labours we are indebted for some of the grandest and most valuable of the North American Conifers.

It is found widely distributed over North-West America, California, and on the Rocky Mountains, varying in heights of from 20 to 30 feet in high exposed situations, to fully 200 feet in sheltered valleys. Though only cultivated in this country since 1827, there are already some noble specimens in our parks and pleasure-grounds, giving abundant proof of its thorough hardiness, the freeness and rapidity of its growth, and the facility with which it accommodates itself to almost every kind of soil. The best examples are, however, found in rich loamy or alluvial soils, and in situations moderately sheltered. It is a magnificent park-tree, with a symmetrical, conical outline, the branches very numerous, and disposed in regular whorls round the trunk. As it bears seed freely, and can now be procured in quantity at a comparatively cheap rate, it will in all probability soon be extensively planted in woods for its timber, which is said to be of superior quality and very durable. The leaves are flat, irregularly set in two rows on the branches, of a bright green above and paler beneath.

Of varieties, the following are worthy the attention of planters of ornamental Conifers:—

Var. *taxifolia*, sometimes called *Tsuga Lindleyana*, is a form found on the Real del Monte Mountains in Mexico, at elevations of from 8000 to 9000 feet. It is also found in Oregon, and is in all probability the tree first discovered by Menzies. The branches are stouter and more erect and the leaves larger than those of the species. It is rarely found growing to greater heights in its native habitats than from 30 to 40 feet. In our pleasure-grounds it forms an exceedingly handsome specimen, quite distinct from the species, and well deserving of being largely introduced among the choicer of the strong-growing Conifers.

Var. *Stairii*.—Since the introduction of the species variegated varieties have occurred from time to time, some of them of great beauty; of these the finest is that which we notice here. It originated at Castle Kennedy, one of the seats of the Earl of Stair, and has within the last few years been distributed under this name. In an article in the 'Garden' of November 23, 1872, Mr D. J. Fish, the well-known horticultural writer, says: "I saw the parent plant of this recently at Castle Kennedy. It is a strikingly beautiful tree even in the autumn, but far more so in the spring, when it is a veritable silver, indeed almost a pure white species. This, unlike some so-called variegations, is not the result of weakness or delicacy in constitution. I had the opportunity of examining several hundreds of these beautiful trees, which in hardiness, rapidity of growth, and vigour of constitution, seem to equal their green parent. There can be little doubt that a brilliant future is in store for this Silver Spruce in our woods and landscapes.

It is impossible to conceive anything more novel and charming than a free-growing Spruce with young shoots almost as white as the *Acer negundo variegatum.*" Whether the anticipations of Mr Fish may be realised, and that the young plants propagated by cuttings and grafts will maintain the character of the parent when grown in other soils and situations, remains to be proved; meanwhile it is evident that the variety is at least a hopeful acquisition, and well worthy of a trial in any collection of fine Conifers.

A. Englemanii (*Engleman's Spruce*).—This species is indigenous to the higher parts of the Rocky Mountains, from New Mexico to the head-waters of the Columbia and Missouri rivers, forming immense forests at elevations of from 8000 to 10,000 feet, and varying in heights from the mere bush of from 2 to 3 feet on the higher summits, to the stately tree of more than 100 feet in the sheltered valleys. It is here of slow growth, but a particularly elegant and distinct-looking tree, quite hardy, and likely to prove an invaluable acquisition to our ornamental evergreen trees. The leaves are light green, somewhat glaucous, thick and bristly, and densely scattered over the shoots. The branches are robust, very numerous, and regularly disposed round the stem from the ground upwards, giving the plant a neat conical form. It seems to prefer a damp, rich soil, and should be planted in a situation sheltered from the blast. In a young state it forms a pretty lawn specimen, and will be valuable as a contrast with the darker-coloured trees in mixed plantations.

A. excelsa (*the Norway Spruce*).—This species, indigenous not only to Norway but to several other countries in the north, both of Europe and Asia, is a lofty evergreen tree, forming in its native habitats vast forests, and attaining heights of from 80 to 100 and even 150 feet, with trunks of from 4 to 5 feet in diameter, the largest specimens being found in sheltered valleys growing in deep, moist, loamy soil. It supplies the wood so much used all over the Continent, as well as in this country, known as *white deal*.

Since its introduction to Britain, more than 300 years ago, it has been extensively planted in forests for its timber, and for producing shelter for other trees of more tender constitutions,—a purpose for which, from its perfect hardiness and close habit of growth, it is peculiarly adapted. No tree has a more striking or characteristic effect on the landscape, whether as seen in groups or interspersed with others in the mixed plantation; while its fine symmetrical outline, regularly and densely arranged branches, and dark-green foliage, render it one of the most effective and desirable of park or pleasure-ground specimens. The cones, which are produced abundantly at a comparatively early stage of its growth, are from 5 to 6 inches long, pendulous, and produced at the tips of the branchlets; when young, they are of a bright red or scarlet colour, and contrast pleasingly with the deep green foliage.

There is a considerable number of very distinct varieties, some of them of great beauty, and nearly all worthy the attention of planters of ornamental Conifers: these include several forms of very dwarf growth, valuable for rockeries, front rows, and small lawns. They should be planted in light, well-drained soils, and in sunny exposures.

Var. *archangelica*.—This is a dwarf variety seldom seen above 4 feet high; the leaves are thick and bristly, and the branches short and much divided into small stiff branchlets. It is distinct, and interesting as a rock-work shrub.

Var. *Clanbrassiliana*, known as Lord Clanbrassil's Spruce, is another dwarf form, never growing higher than about 3 feet, forming a round, close, cushion-like plant, of remarkably slow growth, and well suited for a small lawn or rock-garden.

Var. *findonensis*.—This is a peculiarly handsome plant, said to be of English origin: it differs from the species in having the leaves on the upper or more exposed parts of the shoots tinted with pale yellow, changing as they become matured to a deep brown, while those on the under side are deep green. It is a very distinct and desirable lawn plant, and as it will probably attain something like the height of the parent, it might with advantage be planted in parks and extensive pleasure-grounds.

Var. *Gregoriana* has a bushy, conical habit of growth, the branches very short and abundant, the leaves bristly, and of a bright green colour. It is a neat, dwarf, slow-growing plant, very desirable for any arrangement of small shrubs or trees.

Var. *inverta*.—This is a most distinct variety, with graceful weeping branches, the main stem quite erect, and furnished to the ground. It is of very free growth, and makes a grand lawn specimen plant, in habit of growth suggestive of the *Cedrus deodara*.

Var. *monstrosa*, a curious Araucaria-like form, with long, straggling branches, very sparingly furnished with branchlets: the leaves are thick and bristly, and of a deep green colour. It is an interesting and appropriate plant for a rockery, and, indeed, well worth growing anywhere for its grotesque appearance.

Var. *pyramidalis*, a miniature of the species, with a close, upright habit of growth, forming a superb lawn specimen.

Var. *pygmæa*.—This diminutive plant is the smallest of the varieties, never growing higher than a few inches above the ground. Its habit of growth is dense, and it makes very slow progress, even in the most favourable circumstances.

A. **Menziesii** (*Menzies's Spruce*).—This magnificent species was discovered and sent home by Douglas in 1831 from Northern California, where it occurs over a wide area, growing in damp or marshy soils at altitudes of from 7000 to 9000 feet, and attaining heights of from 80 to 100 feet. It is described by Nuttall as "constituting the principal part of the lofty and dark forest which caps the summit of Cape Disappointment, at the entrance of the Columbia or Oregon."

In our pleasure-grounds it is one of the hardiest of evergreen trees, of free and rapid growth in rich damp soils, and, besides being very ornamental, might be profitably planted in forests for its timber, which, though somewhat coarse-grained, is said to be durable, and useful for a great variety of common purposes.

It forms a stately park specimen, having a symmetrical conical outline, regularly and densely branched from the ground upwards. The leaves are bristly, very sharp pointed, bright green on the upper surface and silvery beneath, giving the tree a shining appearance when the branches are agitated by the wind.

A. **morinda** (*the Himalayan Spruce*), also known as *Khutrow* and *Smithiana*, is a noble tree found on the Himalayas at elevations of from 6000 to 12,000 feet above the level of the sea, and attaining heights of from 80 to 150 feet, with trunks of from 12 to 20 feet in circumference near the ground. It was first sent home in 1818, and has since been widely distributed over the country as an ornamental tree, proving itself perfectly hardy in most districts, and forming a remarkably handsome lawn or park specimen, with a neat conical habit of growth, thickly clothed to the ground with drooping branches. The leaves are profusely scattered over the shoots, from 1 to 2 inches long, and of a deep green colour. It should always be planted in high but well-sheltered situations, and in rich soils, with light porous subsoils : in cold, damp localities, it frequently sufers damage from spring frosts.

A. nigra (*the Black American Spruce*).—This species is indigenous to the coldest regions of Canada and the United States, in similar localities to the White Spruce, occurring in many districts in great abundance, particularly in deep, moist, alluvial soils, and attaining the height of from 50 to 80 feet. The famous American beverage, "Spruce-beer," is manufactured from an infusion of the young shoots of this tree. The wood is strong, light, elastic, and is much used in America in shipbuilding.

It was first introduced into this country in 1700, and has been much valued as an ornamental tree, having a neat, sharply conical habit of growth, very densely furnished with branches. The leaves are thickly scattered over the branches, very short, and of a peculiarly dark-green colour above and bluish-white below, contrasting very effectively with other trees of a lighter shade. It is perfectly hardy here, but will only thrive and form a handsome specimen in cold, moist soils.

A. orientalis (*the Eastern Spruce*), indigenous to mountains in the Crimea, and abundant on the shores of the Black Sea, from whence it was first introduced in 1825, is one of the finest of the species, rising in its native valleys to heights of from 60 to 80 feet, with a straight stem of from 1 to 3 feet in diameter near the ground. It is here perfectly hardy, of somewhat slow growth, but at all times a most beautiful plant, with a broadly conical habit, thickly branched from the ground. The leaves are very short, of a fine dark-green colour, and produced in great profusion. It luxuriates in a damp peaty soil, and does well in any situation slightly sheltered from the force of the wind. For a lawn of small extent, or, indeed, wherever a neat compact evergreen is desirable, no plant can be more strongly recommended.

A. obovata (*the Obovate-coned Spruce*).—This species is indigenous to mountains in Siberia, particularly those of the Altai range, where it is said to occur in great abundance at altitudes of from

2000 to 5000 feet—varying in height, according to soil and exposure, from a dwarf bush to a stately tree of about 100 feet.

It is perfectly hardy here, and is frequently met with in ornamental plantations, forming a remarkably handsome, densely-branched, conical tree, with a general appearance suggestive of the Norway Spruce. The branches are divided into innumerable small branchlets, much more slender than those of that species, and thickly clothed with very short, light-green leaves. It is of slow growth even under the most favourable circumstances, but makes a neat lawn specimen, thriving well in most soils if moderately rich and not too dry.

A. polita (*the Corean Spruce*).—This tree is found in vast forests on high mountains in the peninsula of Corea, attaining heights of from 80 to 100 feet. It is also indigenous to Japan, where it is extensively cultivated for the ornamentation of pleasure-grounds, and from whence seeds were sent home by Mr J. G. Veitch in 1860. From the fact that the name *polita* was first given to a plant which afterwards proved to be identical with *A. morinda*, and that it is still noted in some catalogues as a synonym of that species, doubts have been expressed as to the distinctness of Mr Veitch's introduction. It is, however, totally different, and has a greater affinity, both in style of growth and foliage, to the Norway Spruce than to its Indian congener, or indeed to any of the other species of the genus.

Though as yet only seen here in a young state, and not very extensively distributed, it promises to be one of the handsomest of our ornamental Firs, quite equal to our climate when planted in a moderately sheltered situation, and growing with great vigour in any rich, humid soil. Its habit of growth is broadly conical, the branches very numerous, regularly arranged round the stem, and divided into numerous stiff branchlets amply clothed with long, bristly leaves, of a warm green colour.

A. Pattoniana (*Patton's Spruce*), also known as *Craigiana*, and probably also identical with the plant introduced by Mr Murray, and named *Hookeriana*, was first discovered by Messrs Lewis and Clarke in Upper California, when exploring the sources of the Missouri, and described as one of the most magnificent of the Californian Conifers, growing at altitudes of from 4000 to 6000 feet, and rising to heights of from 200 to 300 feet, with a trunk of from 12 to 14 feet in diameter. It was subsequently discovered by Jeffrey in Northern California, by whom it was first sent to this country, and named in compliment to the late Mr Patton of The Cairnies, Perthshire. In his description of this tree, Mr Jeffrey says: "It is a noble tree, rising to the height of 150 feet, and $13\frac{1}{2}$ feet in circumference, and towering above the rest of the forest; but as it ascends the mountain it gets gradually smaller, until at last it dwindles down into a shrub not more than 4 feet high."

It is here perfectly hardy, of somewhat slow growth, but forms a distinct and remarkably handsome ornamental tree, with slender branches densely clothing the stem to the ground. The leaves are short, very numerous, of a light-green colour on the upper surface, and glaucous on the under. It is found to succeed best in a deep, rich, loamy or peaty soil, not over moist.

ARAUCARIA.

In this genus there are a number of handsome evergreen trees, indigenous to Australia and South America, some of them very lofty, and much valued for their timber, which is durable, easily wrought, and capable of receiving a brilliant polish; while the seeds of several of the species supply an important article of food to the inhabitants of their native regions.

With the single exception of *imbricata*, however, they are too tender for the open air in Britain, though several of the species are elegant and effective as conservatory or lofty greenhouse plants, and as such have long been cultivated, their singularly graceful habits of growth and distinct appearance contrasting well with the other occupants of such structures. Among the finest of these are *excelsa*, known as the Norfolk Island Pine, *Bidwellii*, *Cunninghami*, and *Cookii*, all Australian or South Sea Island species, with the noble *brasiliensis* from the mountains to the north of Rio de Janeiro, with long, lanceolate leaves, and plume-like habit of growth, very nearly but unfortunately not quite hardy, though in very mild winters and particularly favourable situations it sometimes stands out without injury.

A. imbricata (*the Chile Pine*) was introduced into British gardens in 1796 from the Andes of southern Chile, where, as well as on several of the other high mountain-ranges of that country, it occurs in vast forests at altitudes nearly approaching the snow-line, and attaining heights of from 100 to 150 feet.

The wood is reported as strong, very durable, and beautifully veined when polished, and being easily wrought, is used for a great variety of purposes. The large nut-like seeds, which are produced in great abundance, are nutritious and palatable when roasted, and are used as food by the Araucanos and the other Indian tribes who inhabit its native mountains, besides forming an important article of commerce, large quantities being regularly sent down to the markets of the various towns.

It forms a neat, sharp-pointed, conical tree, the main stem perfectly straight, with the stiff horizontal branches arranged round it in regular whorls from the ground upwards, both stem and branches being so densely covered with its strong lancet-like leaves that no animal can climb it or even rub against it without being hurt—hence its native name, *Pehuen*, or "Puzzle Monkey."

Few ornamental trees are more extensively grown or more universally popular; and among the many magnificent forms of the Coniferæ which now so richly adorn the parks and pleasure-grounds of our country, it is undoubtedly one of the finest; unique in its massive grandeur and symmetry of form, and producing, wherever introduced, effects which never fail to arrest the attention and call forth the warmest expressions of admiration from all those who are able to appreciate the noble and picturesque in trees.

In order to the complete development of its beauties, the Araucaria should be grown as a single specimen, with abundance of space on every side to allow its branches to spread out to their fullest extent. In planting, this should be carefully kept in view, as no plant suffers more readily from confinement or contact with other trees—the side shoots either turning to the open side or dying off, to the permanent detriment of the superb conical form so characteristic of the tree. There are few trees more striking for long avenues, either when planted in single rows by themselves or alternately with the lighter-foliaged Conifers—such as *Cedrus deodara, C. atlantica,* or *Abies Albertiana*—its dark sombre green, and stiff formal habit, contrasting admirably with the warmer tints and soft feathery outlines of such species.

It is scarcely necessary to say that it is perfectly hardy in every district in this country; and this, notwithstanding the fact that many specimens of all sizes and ages were damaged, and not a few altogether destroyed, by the exceptionally severe winter of 1860-61, when in some places the thermometer fell to 38° below the freezing-point. A large percentage, however, passed through this severe ordeal scathless; and there can be no doubt that the damage which the others sustained was to be attributed more to their being in low, damp situations, or to some other circumstances equally unfavourable to the proper ripening of their growth, than to any want of hardiness on their part to withstand the frost.

As regards soil, less seems to depend upon its composition than upon its mechanical condition, as fine vigorous specimens are to be found almost everywhere, the failures being for the most part in cold, stiff clays, with wet, impervious subsoils. Where such is the state of the ground, the soil should be removed to the depth of 3 feet, the bottom well loosened up and drained, and the pit fitted up with good porous loam or rich compost. Where the situation is very damp, and complete drainage impossible, it is a good plan to raise the plant to the surface, banking the new soil round it. This is resorted to very frequently in such circumstances with complete success, the long under branches bending down and covering the bank in such a way as to modify what might be considered a somewhat unsightly object.

Var. *variegata* has its leaves more or less prominently tinged with a straw-coloured variegation: though interesting as a variety, it cannot be ranked among the finest of the variegated Conifers.

BIOTA (THE EASTERN ARBOR VITÆ).

The handsome evergreen shrubs which constitute this group are for the most part indigenous to China, Japan, Tartary, and Northern India. They were originally associated with the *Thujas*, of which they formed the second section, but are now separated into a distinct genus under the sectional name by which they were distinguished from the American or Western species.

All the sorts in cultivation are found to grow freely in ordinary garden soils, but prefer such as are dry, with the subsoils warm and porous; and though most of them are fully equal to the rigours of our winters in the open air, they succeed best where they are protected from the full force of violent winds.

The following species and varieties are among the most distinct and ornamental :—

B. orientalis (*the Chinese Arbor vitæ*), found wild in great abundance in mountainous districts in China and Japan, growing to heights of from 20 to 30 feet, was introduced into Britain about 1752. With a general resemblance to the American Arbor vitæ, it is readily distinguished not only by the peculiar form of its cones, but by its much more sharply conical, almost columnar, habit of growth, compact, erect branches, and dense, flat branchlets. The branchlets have a warm green colour in summer, changing, particularly if the plant is growing in an exposed situation, to a brown tint on the approach of winter.

Ever since its introduction, this beautiful shrub has been one of the most conspicuous and highly valued ornaments of our gardens and pleasure-grounds; and, notwithstanding the many brilliant acquisitions to the list of hardy Conifers during the last twenty years, it has even yet few rivals for real elegance and symmetry of form, and is still, as it richly deserves to be, extensively planted in the most choice collections.

Though quite hardy, and of free growth in most

districts, the finest specimens are invariably found where the soil is a deep light loam, the land well drained, the situation airy but sheltered, and plenty of space allowed for the branches being freely developed on every side.

Like many other plants of a wide geographical range, the Chinese Arbor vitæ is prolific in varieties: of these, the following are among the most useful for decorative purposes :—

Var. *aurea*, or, as it is popularly called, Golden globe, is so very different in its style of growth from the parent, that at first sight it is difficult to realise the fact that it is not itself a distinct species, but a mere sport originating from seed saved from *orientalis*. This lovely little shrub is rarely seen higher than about 3½ feet, and has an almost completely globular habit of growth, the branches so abundant and so dense as to suggest the idea of solidity. In spring and early summer the young branchlets have a most brilliant golden hue, gradually changing as they become matured to the lightest green. From its neat, close style of growth, pleasing colour, and hardiness, it is well adapted for planting in terrace or flower-garden vases, in centres of geometric figures, or on small lawns. It should have a sunny but moderately sheltered situation, and delights in a rich, dry soil.

Var. *globosa* has the peculiar, compact, globular

habit of growth of *aurea*, but with dark-green instead of golden branchlets: it is a fine companion plant to that variety.

Var. *compacta* is a very elegant, sharply conical, dwarf bush, with abundance of branches densely clothing the stem from the ground upwards.

Var. *elegantissima* has the compact, columnar form of the species, and in summer its green colour. It changes, however, in winter to a bright reddish-brown, the young branchlets in spring being more or less tipped with gold.

Var. *semper-aurescens* is a new and very interesting form, of Continental origin, resembling the preceding in general appearance, but with a finer golden variegation, which it retains with more or less brilliancy all over the year. Though comparatively little known here, this pretty plant will doubtless soon be widely distributed.

Var. *freneloides* or *gracilis* is another singularly beautiful and distinct variety, found on mountains in Northern India, and, like the species, of an erect, closely-branched habit of growth, but much more slender in all its parts. It is very distinct, quite hardy in sheltered situations, and deservedly very popular.

Var. *pyramidalis* or *tartarica* is a distinct and handsome form, with more robust branches, dense branchlets, and a more compressed, columnar style of growth than the species, and makes a superb lawn specimen.

B. filiformis or pendula (*the Thread-branched Arbor vitæ*) is found wild in high mountain valleys in Japan, particularly on the Hakone ranges, forming a bush varying in height, according to soil and situation, from 10 to 20 feet. From its handsome appearance and freeness of growth, it is highly valued and extensively grown in Japan, and all over China, as an ornamental shrub.

It was first introduced into British gardens early in the present century; and though very properly classed among hardy Conifers, it will only succeed in well-sheltered localities. A rich, deep, loamy soil, rather light than heavy, with the subsoil either naturally dry and porous or thoroughly drained, is also an essential condition to its wellbeing, and must be supplied, if even a tolerably vigorous and handsome specimen is to be obtained.

In favourable circumstances, this curious and interesting plant forms a straight-stemmed, bushy-headed shrub or miniature tree, with long whip-cord-like pendent branches, rather sparingly clothed with close, small, scale-like leaves clinging to the stem. The branchlets are numerous, and disposed at irregular intervals on the branches. In summer it has a bright green colour, changing in winter to a dark brown.

B. japonica (*the Japan Arbor vitæ*).—As its name implies, this species is a native of Japan, where it forms a bushy shrub, rarely exceeding 10 feet in height. It was first sent home to this country in 1860, and has proved itself to be quite hardy, and to grow freely in similar circumstances to *orientalis* and its varieties, which it so much resembles that some botanists have doubted its claim to rank as a distinct species. Whether a species or variety, however, there can be no question of its being a great acquisition to our hardy Biotas, nor of its worthiness of admission to the most choice collection of ornamental Conifers. In habit of growth it is more broadly conical than *orientalis*, but equally dense, the branchlets more compressed and fan-like, and forms a symmetrical specimen shrub.

B. meldensis (*the French Hybrid Arbor vitæ*).—This distinct and interesting plant originated some years ago in Meaux, in France, from whence it was distributed as a hybrid between the Arbor vitæ and red Cedar.

Its hybrid origin, however, has been doubted by some of the highest authorities on such subjects, and it is now generally believed to be a seminal sport from some of the Biotas. It is quite hardy here in sheltered situations, and forms an erect, somewhat dwarf, bushy shrub, with slender, slightly drooping branches, disposed irregularly but very abundantly over the stem. In summer it is of a light, slightly glaucous-green colour, changing in winter to a red or reddish brown. Where this succeeds, it forms a pretty lawn plant, and is useful in the mixed shrubbery as a contrast to the commoner tints of foliage.

CEDRUS (THE CEDAR.)

In this genus we have a group of three evergreen trees of surpassing beauty, very hardy, and so universally admired that they are to be found occupying the most prominent places in almost every Pinetum and ornamental plantation throughout the country, each of the three forms having a distinct habit of growth, and a general appearance peculiar to itself. Few trees are less fastidious in regard to soil, growing with equal luxuriance among the loams and clays of the granite and trap and those of the sandstone systems; while in the lias and chalk of England they are most vigorous, and become magnificent specimens: they delight, however, in a rich soil with a cool subsoil, but well drained, as they are most impatient of water at the roots.

The leaves resemble those of the Spruce Firs, being needle-shaped and short, but are disposed on the mature branches in tufts, varying in number from five to about thirty, and scattered singly only on the young shoots.

C. libani (*the Cedar of Lebanon*).—This magnificent and well-known tree is indigenous to Mount

Lebanon and other mountains in Asia, and is found distributed more or less abundantly over a wide range.

It was first introduced into Britain nearly 200 years ago, and many of the original trees are to be met with in our parks and pleasure-grounds which have attained a large size, and are said to equal, if not surpass, any now to be found on Mount Lebanon itself. It grows to heights of from 70 to 80 feet, with trunks varying from 30 to 40 feet in circumference near the ground; becoming, when old, a flat-headed tree, irregularly branched, and when standing alone, very broad in proportion to its height. The branches, which spread out horizontally from the stem, are so densely furnished with shoots and foliage as to suggest the idea of a series of solid shelves or stages. The leaves are in alternate tufts of about thirty in number, about 1 inch long, stiff and sharp-pointed, and of a grassy-green colour.

Apart from the interesting associations connected with this tree as the "goodly cedar" whose timber was employed in the construction of Solomon's Temple, and otherwise so frequently alluded to, and its various characteristics so graphically described in the emblems and metaphors of Holy Writ, it has peculiar claims upon the attention of the decorative planter, being beyond all question one of the finest of our park trees, thriving well in almost every situation, handsome at all stages of its growth, and, when at maturity, unrivalled for its stately grandeur and picturesque beauty, bringing forcibly to remembrance the striking words of Ezekiel: "Behold, the Assyrian was a cedar in Lebanon with fair branches, and with a shadowing shroud, and of an high stature; and his top was among the thick boughs."

Though sometimes planted in groups by themselves, or associated with other trees in mixed plantations, the Cedar of Lebanon is seen to the best advantage standing alone, with no other object near to prevent the development of its long, spreading branches, or to obstruct the view of its bold, irregular outlines.

It prefers a moist, deep soil, and, it is scarcely necessary to say, is so hardy that it is never injured by the severest frosts.

C. deodara (*the Deodar or Indian Cedar*), found in vast forests in the ravines of the Northern Himalayas, at elevations of from 5000 to 12,000 feet above the sea-level, and rising to heights of from 150 to 200 feet, with trunks of from 20 to 30 feet in circumference. Since its introduction in 1822, this singularly beautiful tree has been most extensively distributed over every district in Britain, proving perfectly hardy, and equal to our severest winters, if planted in tolerably rich, well-drained land and sheltered from the full force of violent winds.

If the Cedar of Lebanon may with truth be said to be the most majestic and picturesque of park trees, the Deodar may with equal truth be classed as at least the most graceful. Its symmetrical, broadly conical habit, densely clothed to the ground with long, pendulous branches, along with its beautiful glaucous colour, never fails to please, and to secure for it a place in the most select collections. From its peculiarly handsome habit of growth, it is most admirably adapted for planting singly on lawns, and is unsurpassed as an avenue tree, if allowed plenty of space for the development of its branches. It is one of the few Conifers that may be freely and in many cases advantageously pruned, an operation which we recommend being performed in spring, before the growth commences.

The wood of this tree is considered superior in quality to that of either *libani* or *atlantica*, being much more resinous, closer grained, and consequently more durable; and is so much valued by the Hindoos that they use it in the building of their temples, as well as for other purposes where durability and strength are objects of importance. By a striking coincidence, they call it "Devadara," the Tree of God; and the same term is applied by the Psalmist in the 80th Psalm to the Cedar of Lebanon, the "goodly Cedars" being literally "Cedars of God," —a fact which, along with the superiority of its timber, has led some writers to suppose that it was the Deodar and not the Cedar of Lebanon that was used as beams in the Temple of Jerusalem. This, however, it need scarcely be said, is extremely improbable, seeing that the tree has never been discovered nearer to Palestine than the Himalayas.

The following varieties are very distinct and ornamental; but, from their requiring to be propagated by grafts or cuttings, they are as yet comparatively scarce, and not so frequently met with as the species.

Var. *robusta*.—In this form the leaves are longer and much thicker than those of the species, which it otherwise closely resembles: it is a remarkably distinct and desirable lawn plant.

Var. *viridis*.—This differs from the parent in having smaller leaves, which, instead of being glaucous, are of a bright green tint: it is a well-marked and extremely interesting plant.

Var. *verticillata glauca*.—A robust-growing variety, with larger leaves than the species, and remarkable for its prominent glaucous colour: it is worthy of a place among the finest shrubs.

C. atlantica (*the Mount Atlas Cedar*), also called *argentea africana* and *elegans*, is indigenous to the highest mountains of Barbary, and most abundant on the great Atlas range, at elevations of from 5000 to 10,000 feet above the sea, and described as a noble and picturesque tree, rising to heights of from 70 to 100 feet. In general appearance it presents a combination of the characteristics of the other two forms, having the tall, straight, tapering habit of the Deodar on the one hand, with the stiff horizontal branches of the Cedar of Lebanon on the other. Since its introduction into this country about the year 1845, and particularly during the last dozen of years, it has been extensively planted as an ornamental tree, and has proved itself to be one of the hardiest and most distinct of our recently introduced Conifers—its shining, silvery foliage, and dense, conical habit of growth, contrasting well with most other trees of the Fir or Pine tribe.

Though as yet for the most part planted only for ornament, this species has many qualities to recommend it to the attention of those interested in timber-trees. Its perfect hardiness, and ability to stand uninjured in the most exposed situations; the rapidity of its growth, in favourable circumstances rivalling that of the Larch; and the fact that the wood is equal in value to that of any of the other coniferous trees, must, as soon as seeds can be got cheap enough and in sufficient quantity, secure its being extensively planted as a profitable timber-tree. Like the other species, it succeeds best in rich soils, and delights in a stiff damp subsoil.

CRYPTOMERIA (THE JAPAN CEDAR.)

This is a small genus of evergreen trees indigenous to Japan and China, occurring in great abundance in a wild state, and extensively cultivated in both countries as ornamental trees. The wood is soft and light, and much valued by the Japanese for the finer kinds of cabinet-work. Though by no means to be classed among our hardiest trees, they are found to succeed well in sheltered localities, and there are innumerable fine specimens to be seen in almost every district of the country.

C. japonica (*the Japan Cedar*).—This elegant species, for which we are indebted to the indefatigable and successful botanical collector, Mr R. Fortune, who first sent home seed in 1844, is found wild, not only in Japan, but in great abundance over an extensive area in the north of China, growing most luxuriantly in damp soils with a hard and rocky substratum, at elevations of from 500 to 1200 feet above the sea-level, and rising to heights of from 60 to 100 feet. It is here a distinct-looking, symmetrical tree, with a sharp, conical habit of growth, densely branched from the ground, making an admirable lawn or avenue specimen. The leaves, which are short and needle-shaped, cover the shoots very thickly, and are of a bright green colour. A sheltered situation is indispensable to its wellbeing;

and though it grows luxuriantly in rich, deep soils, with cool, damp subsoils—in order to secure the wood being early ripened in autumn, it should only be planted in such as are light and porous. There are several varieties of this interesting tree in cultivation, the most distinct and useful of which are:—

Var. *Lobbii*, or, as it is sometimes called, *viridis*, has a dwarfer and much more compact style of growth than the species, and the foliage of a brighter green. It is a very pretty shrub or miniature tree for small lawns or gardens. Like the species, it stands well in moderate shelter, and indeed seems of stronger constitution.

Var. *nana* is a very dwarf variety, forming a curious flat-headed bush from 2 to 3 feet in height. Though not very graceful, it is interesting enough for planting on or about rockeries, and is frequently found in such situations.

Var. *araucarioides*, another dwarf variety, quite distinct, but resembling *Lobbii* in its style of growth. It is well worthy of a place among choice dwarf shrubs.

C. elegans *(the Elegant Cryptomeria)*, introduced in 1863 by Mr J. G. Veitch from Japan, where it forms a shrub-like tree of about 30 feet in height, growing naturally in high but sheltered places, in rich soils, with warm, porous subsoils. It has withstood the winters of the past few years with perfect safety, and is found to grow freely in almost every variety of garden soil, if well drained and protected from violent winds.

It is unquestionably a valuable acquisition, and doubtless, as its merits become better known, will be largely cultivated.

The habit of growth is broadly conical, the branches thick and massive, and the branchlets drooping at the extremities, giving the plant a weeping appearance. The leaves are much longer than those of *japonica*, of a light green in summer, changing in autumn and winter to a brownish-green or bronze tint. It is then very striking and distinct-looking, and invaluable as a contrast to most other shrubs.

CEPHALOTAXUS (THE CLUSTER-FRUITED YEW).

Though botanically different from their near allies the Yews, the species which constitute this genus have a general appearance suggestive of that group, and are associated with them in the coniferous sub-family *Taxineæ*.

The few species in cultivation are natives of Japan and China, where they are found at high altitudes, forming medium-sized trees, valued for their timber, and extensively cultivated in the gardens and pleasure-grounds of both countries as decorative trees.

If less hardy and somewhat more fastidious in regard to soils and situations than the Yews, all the species are quite handsome enough to justify their being recommended for trial wherever a well-sheltered locality and a deep, rich, loamy soil can be made available; and where they do succeed they are objects of great beauty, their warm green tints blending most pleasingly with the darker green of the common Yew and its varieties.

Like the Yews, they are peculiarly adapted for shady situations, where, if the soil is moderately moist, and each plant sufficiently isolated to admit of the branches being freely developed on every side, they rarely fail to form neat and attractive specimens.

C. Fortuneii *(Mr Fortune's Cephalotaxus)*.—This is one of the many valuable accessions to our list of hardy plants for which we are indebted to the distinguished collector whose name it bears, and by whom seeds were first sent to this country in 1848. It is found wild in several districts in Japan and the north of China, chiefly in high but sheltered valleys, and attaining heights of from 40 to 60 feet.

Though of slow growth, and thriving here only in well-sheltered localities, it is in favourable circumstances an interesting and distinct-looking shrub, of a broadly conical shape, sparingly furnished with long slender branches, divided into numerous branchlets slightly drooping at the points: the leaves are from 2 to 3 inches long, of a linear-lanceolate form, the upper surface bright glossy green and the under dimly glaucous. There is a very desirable variety of this species obtained from seed, and recently introduced into cultivation, which has been named *F. robusta*, only differing, however, in its denser and more vigorous habit of growth.

C. drupacea *(the Plum-fruited Cephalotaxus)*, another of Mr Fortune's fine introductions, and found wild in similar localities in Japan and the north of China as the preceding, is in its native habitats a small bushy tree, rarely exceeding 30 feet in height.

This sort was at first believed to be the female form of *Fortuncii*, and was distributed under the name *F. fœmina;* but as both sorts have since been found to be fruit-bearers, the propriety of ranking it as a distinct species is sufficiently obvious.

As yet it has only been seen here as a moderate-sized shrub, with a close, conical habit of growth, the branches much divided into small branchlets, profusely clothed with yew-like leaves from 1 to 1½ inch long, of a yellowish-green colour on the upper surface, and faintly silvery on the under.

Being much hardier and more accommodating in regard to soils and situations than *Fortuncii*, it is more frequently met with in collections; and though lacking the distinct tropical aspect peculiar to that species, it is on the whole a more useful outdoor ornamental shrub, and makes a handsome lawn specimen in ordinary soils, if moderately sheltered.

C. pedunculata (*the Long-stalked Cephalotaxus*).—This fine species, still better known under its original name *Taxus Harringtonia* (the Earl of Harrington's Yew), was first introduced into British gardens in 1837 from Japan, where in high sheltered valleys it forms a broad, bushy shrub, frequently attaining heights of from 20 to 25 feet. It is the *Inukaja*, or "Wild Yew," of the Japanese; and is by them extensively employed as a hedge-plant as well as for the decoration of their gardens and pleasure-grounds.

In this country it forms a dense, spreading bush, very abundantly branched, the branchlets short and slightly pendent: the leaves are from $1\frac{1}{2}$ to $2\frac{1}{2}$ inches long, closely set on the shoots in two rows; of a glossy light-green tint on the upper surface, and having a silvery glaucous band underneath on each side of the midrib.

Like the other species of Cephalotaxus, a mild or well-sheltered locality is indispensable to its successful cultivation; and in such, provided the soil is rich and the subsoil cool without being damp, it is a most beautiful and interesting plant, and well worthy of a prominent place in any collection of the choicer Conifers.

CHAMÆCYPARIS SPHÆROIDEA (THE GROUND CYPRESS).

This fine evergreen tree, the "White Cedar" of the Americans, and the only species of the genus of undoubted hardiness in our climate, is perhaps still best known in collections under its original and certainly most expressive name, *Cupressus Thyoides*, or Thuja-like Cypress. It is a native of Canada and the United States, where, particularly in the maritime districts, it has a wide distribution, covering vast tracts of low, swampy land, and growing with a straight, tapering trunk, to a height of some 70 or 80 feet.

The wood is light, fine-grained, and easily wrought, and is said to resist the influence of the weather better, and to be more durable for outdoor purposes, than that of any of the other American Pines.

Though introduced into this country so long ago as 1736, and since then freely planted for pleasure-ground decoration, it has rarely if ever been found to grow higher than a moderate-sized shrub, and it is only as such that it can be recommended to cultivators. In congenial circumstances, however, it is exceedingly handsome, and forms a neat specimen, very hardy, and by no means fastidious in regard to soil, provided that it is deep and rich, with the subsoil stiff and cool, rather than gritty and porous. We are convinced that many of the failures in the cultivation of this very beautiful species are to be attributed more to its being planted in poor light soils than to any want of adaptability on its part to our climate.

In general appearance it resembles some of the Arbor vitæs or Cypresses, or rather combines the characters of both, those of the latter being most conspicuous. Its habit of growth is close, bushy, and sharply conical, the abundant branches being divided into numerous short, twiggy, fan-shaped branchlets, densely covered with short, scale-like leaves of a glaucous-green hue.

The following varieties, which have occurred from time to time among seedlings, are really distinct, and deservedly very popular among admirers of fine shrubs:—

Var. *atrovirens*.—This differs from the parent only in the colour of its foliage, which, instead of being glaucous, is bright green.

Var. *glauca* is dwarfer and more compact in its habit of growth, the foliage of a more decidedly silvery-glaucous colour, which it retains throughout the year.

Var. *variegata*.—This, by far the finest of the varieties, and with some cultivators a greater favourite than the species itself, has its green branches freely intermixed with a golden-yellow variegation. It is one of the prettiest of the variegated Conifers, and, having the neat compact habit of its parent, forms an exceedingly beautiful lawn specimen plant.

CUPRESSUS (THE CYPRESS).

The hardy species of this genus form a group of surpassingly handsome evergreen shrubs or bushy trees, natives of Europe, Asia, and America; and though, with the exception of *sempervirens*, which has been known in British gardens for upwards of 300 years, they are of comparatively recent introduction, no family of plants has been more extensively used in the decoration of our parks and pleasure-grounds. Their distinct, symmetrical style of growth, and beautiful tints of colour, produce the most charming effects, whether in combination with shrubs and trees of other genera, massed by themselves, or planted, as they very often are, singly as picture trees on lawns and flower-gardens.

Though all the sorts prefer a light, loamy soil, they are found to do well and form fine specimens in almost every variety of garden soil, if the ground is thoroughly drained, and the subsoil warm and moderately porous. In every case it is indispensable that they should be so sheltered as to protect them

D

from violent winds, from which, in common with many other perfectly hardy but densely-branched evergreens, they are extremely liable to suffer damage.

All the really hardy species are worthy of being largely represented, even in a limited collection of ornamental shrubs; and while we have noted only such, we may remark that there are others equally handsome, though rather tender for many districts, but which may be planted in the more favoured localities with a prospect of success.

C. sempervirens (*the Upright Cypress*).—This fine evergreen tree is found widely distributed over Southern Europe, Greece, Turkey, Persia, and Asia Minor, and was introduced into Britain in 1540.

Though rarely seen higher in our gardens than from 20 to 30 feet, it rises in its native habitats to heights of from 50 to 80 feet; and in Italy, where it is extensively planted both as an ornamental and as a timber tree, specimens are to be found above 100 feet.

In habit of growth and general appearance, the upright Cypress resembles the Lombardy Poplar; having a straight, spire-like form, the side branches growing erect and close to the stem, the whole plant densely clothed with small, scale-like foliage, of a dark, sombre green colour. The wood has long been famed for its hardness and durability; and being fragrant, beautifully veined, and capable of receiving a high polish, is fitted for the finest cabinet-work.

As an ornamental shrub, it is here much admired, and but for its unfortunate tendency to grow late in autumn, particularly while young, and consequently to suffer injury from early frosts, it would undoubtedly be met with more frequently. If planted, however, in a dry porous soil, and in an airy but well-sheltered situation, it will be found after a few years to ripen its growths earlier in the season, and in most cases to form an attractive specimen.

Var. *horizontalis* differs from the species in being much dwarfer, and in its branches being spread out in a horizontal direction: it is equally hardy, and worthy of a place in a collection of ornamental shrubs.

C. Lawsoniana (*Lawson's Cypress*), indigenous to Northern California, was first discovered by Mr Jeffrey, and subsequently by Mr William Murray, growing in the Shasta and Scots valleys, by whom seeds were sent home in the autumn of 1855. It was named after the Messrs Lawson, nurserymen, Edinburgh.

The character and appearance of this tree, as seen in its native wilds, is thus graphically described by Mr Murray: "This was the handsomest tree seen in the whole expedition. It was found on the banks of a stream in a valley in the mountains, is about 100 feet high and 2 feet in diameter. The foliage is most delicate and graceful; the branches bend upwards at the end like a spruce, and hang down at the tip like an ostrich feather. The top shoot droops like a Deodar. The timber is good, clean, and workable."

In our pleasure-grounds it is a rapid-growing plume-like shrub of a symmetrical conical form, abundantly clothed with branches; the dense branchlets slender and drooping, and of a light-green, sometimes slightly glaucous, colour. The cones, which even on very young plants are produced in great profusion, contain very frequently fertile seeds, and in early summer add to the attractions of the plant by their beautiful golden, and in some varieties purple, colour.

Of a plant so widely distributed, and so extensively planted in almost every shrubbery and park in the country, it seems unnecessary to say much here by way of commendation. Few introductions of recent years possess such a combination of those qualities which are so desirable in an outdoor ornamental evergreen. Its thorough hardiness, freeness of growth in almost all soils if moderately dry and the subsoil porous, along with its singularly graceful habit, must always insure its admission into the most limited collection of decorative trees and shrubs. It is most assuredly one of the finest of our lawn trees, and no plant is better adapted for small villa gardens, where, from its close compact habit, it economises space, and harmonises well with the building, to which it must necessarily be closely planted. We quote the following remarks regarding this tree from an able paper which recently appeared in the 'Garden:' "We can as yet know little of the noble aspect it will assume when it shall attain its full dimensions of 100 feet in height. It will be for the children of those who are fortunate enough to secure some of the largest and finest specimens now obtainable in England or Scotland, to see this grand addition to our recent acclimatised coniferous plants in all its magnificence."

From a considerable number of seminal varieties and sports which have occurred from time to time since its introduction, we note the following as really distinct, and deserving the attention of cultivators: they are all as hardy, of as free growth, and as ornamental, as the parent :—

Var. *argentea*.—This very beautiful plant differs from the species in the branches being of a silvery glaucous tint, and is valuable as a contrast to the other trees and shrubs of a darker-coloured foliage.

Var. *nana* is a very dwarf-growing variety, forming a neat globular bush, densely branched to the ground. It is quite a gem in its way, and well suited for vases or small garden-beds; its foliage is bright green.

Var. *erecta viridis* is a remarkably distinct and elegant form, with a close columnar habit of growth, the foliage very deep green. In a notice of this plant which appeared in the 'Gardeners' Chronicle' a few years ago, the writer justly remarks that it is "quite unapproached for symmetry and beauty by any other plant we know, while the slender ramifications of its close-set compact branches and branchlets give it a degree of refinement which is not seen in any other variety of this grand hardy species. . . . It is utterly unaffected, both as to vitality and hue, by the severest frosts."

Var. *argentea variegata*.—The branchlets of this variety are prominently blotched with a bright, silvery variegation: it is a very effective lawn plant, and ought to be extensively cultivated.

Var. *aurea variegata* has its green branchlets intermixed with those of a bright golden colour, like the preceding, to which it is a fine companion: it is a wonderfully beautiful plant. The variegated branchlets of both, however, are somewhat impatient of frost, and they should only be planted in well-sheltered situations.

C. macrocarpa (*the Large-fruited Cypress*).—This

fine species was found by Hartweg on the hills of Monteroy, in Upper California, growing in great abundance, and rising to heights of from 50 to 70 feet, with a circumference of about 9 feet, and was by him sent home in 1847.

It is here one of the most beautiful of ornamental evergreens, very hardy, and of remarkably rapid growth, forming a compact, cone-shaped specimen, thickly clothed with heavy massive branches and foliage of a bright grassy-green colour. Though quite equal to our climate as far as frost is concerned, it is most impatient of exposure to high winds, which not only seriously damage the foliage, but, from its density, act so powerfully upon it as frequently to tear it out of the ground. A well-sheltered situation should therefore always be chosen if a fine specimen is desired. Any rich garden soil will be found suitable, if well drained, and the subsoil warm and porous.

Var. *Lambertiana*.—This form, introduced by Mr Lambert a few years earlier, though very distinct in its habit of growth, and long considered a distinct species, is now ascertained to be only a variety,— cuttings from old plants of *macrocarpa* generally, though not always, assuming the broad, horizontal-branched, flat-headed habit of *Lambertiana*, exactly corresponding to that of the mature trees on the Californian mountains, which are described as having a "far-spreading, branching, flat top, like a full-grown Cedar of Lebanon." We have therefore in *Lambertiana* a miniature old tree, and in *macrocarpa* the same tree in its young or undeveloped state. Whether this theory is right or wrong, however, we have no hesitation in recommending both sorts as superb and effective lawn plants,—the one so different in outline and style of growth from the other, that they may be planted side by side with the happiest results.

C. Goveniana (*Gowens's Cypress*).—This species, named in compliment to J. R. Gowen, Esq., late Secretary to the Horticultural Society of London, is another introduction by Mr Hartweg, who found it on the western slopes of the mountains of Monteroy, in Upper California, and sent home seeds in 1847, describing it as a dense bush from 6 to 10 feet high.

It is here a pretty evergreen shrub, of a broad conical habit of growth, clothed to the ground with slender feathery branches, drooping gracefully at the extremities, and of light-green colour, attractive at all times, but specially so when in summer it is covered with its bright golden catkins, which it produces at a very early period of its growth. In exposed and wet situations it is liable to suffer from frost, but is found to do well in most districts when planted in a dry gravelly or sandy soil, and well sheltered from violent winds. It is somewhat remarkable that under favourable conditions it grows with great rapidity, and that there are already in this country specimens fully double the height named as the maximum in its native habitats by its discoverer, who in all probability took his notes from plants growing in exposed places or in thin poor soil.

C. M'Nabiana (*M'Nab's Cypress*).—A very beautiful species, named in compliment to James M'Nab, Esq., the urbane and talented Curator of the Royal Botanic Gardens, Edinburgh. It was first discovered by Jeffrey on the Shasta Mountains, in Northern California, at an elevation of 5000 feet, and rarely rising higher than a bush of about 10 feet. Seeds were first sent home by Murray in 1855.

It is here a neat, compact, conical shrub, with fine glaucous foliage, and makes a distinct and interesting specimen for a lawn, particularly useful where an effective dwarf bush is required. It seems to do well in almost all kinds of soil, but thrives to perfection in such as are deep and of a light porous character.

C. nutkaensis (*Nootka Sound Cypress*) is found wild in great abundance about Nootka Sound, and in various other localities on the north-west coast of North America. It was first introduced into

Britain in 1851, and distributed under the name of *Thujopsis borealis*. In its native habitats it forms a straight-stemmed, beautiful tree, of from 80 to 100 feet in height.

It is here a conical, bushy tree, of a most symmetrical form, densely furnished from the ground with spreading branches, much divided into graceful, plume-like, drooping branchlets.

The leaves, which are short, imbricated, and thickly set on the branchlets, are of a dark glossy-green colour, and sometimes, particularly while the plants are young, faintly glaucous on the under side.

Of its merits as an ornamental tree it is impossible to speak too highly. Its distinct, handsome appearance, freeness of growth in almost every variety of soil, and thorough hardiness in our severest winters and coldest situations, mark it as one of the most valuable acquisitions of recent years, and have already secured for it extensive admission into the most select shrubberies and pleasure-grounds. Though a most effective plant in mixed groups or in avenue rows, it is always seen to the best advantage when planted as a single specimen on a lawn, or any situation where, standing alone and free from contact with other plants, it is allowed to develop its beauties in form and colour to their fullest extent. It is found to grow most freely where the soil is a peaty or rich fibry loam, and rather moist than dry;

and while shelter from frost is altogether unnecessary, it should not be exposed to violent winds.

Like many of its congeners among the Cypresses and Thujas, this superb tree is variable in its appearance, and many distinct varieties may be detected among seedlings. Of these the most interesting are—

Var. *glauca*.—Very similar to the species in every other respect but in colour of foliage, which is of a warmer, slightly glaucous green: it is a beautiful and desirable form.

Var. *variegata* has its branchlets prettily tinted with a silvery or straw-coloured variegation: it is a distinct but as yet a scarce variety.

FITZ-ROYA PATAGONICA (THE PATAGONIAN FITZ-ROYA).

This very distinct and handsome species, the only one of the genus, as far as we know, in cultivation, is a native of high mountain-ranges in Patagonia, from whence it was first sent to this country in 1846, and named by Dr Hooker, in compliment to its discoverer, Captain Fitz-Roy. In its native habitats it is a graceful tree, growing in sheltered valleys to heights of from 50 to 100 feet, but dwarfing down to a small bush as it approaches the borders of perpetual snow.

Although found naturally at high elevations, it is here liable to suffer damage from spring frosts, and can scarcely be recommended as suitable for every locality. There are many places, however, where it succeeds; and it is ornamental enough to be worthy of a trial wherever good shelter, a northern aspect, and a dry rich soil are available. It has a dense, bushy habit of growth, the branches abundantly clothed with leaves, somewhat slender and drooping at the extremities.

JUNIPERUS (THE JUNIPER).

The shrubs and trees which form this large and important section of the Coniferæ are for the most part natives of the temperate and colder regions of Europe, Asia, Africa, and America, and are with few exceptions thoroughly hardy, and easily cultivated in Britain.

All the known species are evergreen, and though very varied in stature and style of growth, the great majority of them are valuable decorative plants, and as such are extensively planted in our pleasure-grounds, where, though preferring a deep sandy or gravelly loam, and an elevated rather than a low confined situation, they are found to succeed well in ordinary soils, if well drained.

The Junipers are readily distinguished from their

allies the Cypresses, which in general appearance they very much resemble, by their cones—the scales of which, instead of being hard and dry, are soft and fleshy, like berries—as well as by that peculiar flavour for which they are used in the manufacture of gin and Hollands, and which, though strongest in the fruit, exists to a greater or less extent in every part of the plant.

The wood of all the species, besides being of a beautiful colour, is light, fragrant, and very durable, fitting it for the finest cabinet-work; and that of *bermudiana*, a grand but unfortunately tender species, is the pencil Cedar, well known all over the world.

From a great array of really fine ornamental species and varieties we cull the following, as specially worthy the attention of planters of decorative trees and shrubs:—

J. communis (*the Common Juniper*), found wild in all soils and situations in Scotland, England, and Ireland, as well as over a wide range in continental Europe and Asia, is a bushy shrub, growing to heights of from 10 to 15 feet in good soils and sheltered places, but in high sterile exposures assuming the form of a mere trailing bush, rising only a few inches above the ground.

In cultivation it is a pretty plant, with a close, bushy habit of growth, the branches spreading, rather rigid, with numerous branchlets clothed with linear, sharp-pointed leaves. It is interesting as a distinct variety in a shrubbery or collection of Conifers, but is mostly used for planting in masses for game-cover, for rockeries, or for clothing steep gravelly banks where the soil is scanty and poor.

Of this species the following varieties are old established favourites:—

Var. *suecica*, known as the Swedish Juniper, but also found abundantly in several other countries of Northern Europe, has a narrow, compact, conical habit of growth, with longer leaves and a brighter green colour than the species, and rises in sheltered situations to heights of from 15 to 20 feet.

Var. *hibernica*, the Irish Juniper, found on mountains in Ireland, has short glaucous leaves, and a more compressed, sharply conical form than the Swedish variety. It is a remarkably elegant plant—well suited for a lawn specimen, for associating with the Irish yew in avenues, or for geometric gardens.

Var. *hibernica compressa* is a dwarfer and more compact plant than the preceding, which it otherwise very much resembles.

J. oxycedrus (*the Prickly Juniper*), found widely distributed and in great abundance on mountains in France and several other countries in Southern Europe, growing to heights of from 10 to 12 feet, was first introduced into Britain about 1739. It is a well-known hardy shrub, with an upright conical habit of growth, the branchlets rather open and slightly pendulous at the extremities; the leaves are somewhat larger than those of the common Juniper —stiff, sharp-pointed, and of a warm green colour. It is a very desirable plant in a collection of ornamental shrubs, and thrives well if planted in a dry soil, and moderately sheltered.

J. virginiana (*the Red Cedar*) is a native of the United States of America, where it occurs over an immense area, and in the greatest abundance, rising to heights of from 40 to 50 feet, with a diameter near the ground of 1½ foot. The local name refers to the beautiful red heart-wood of the tree, which, being durable and susceptible of a fine polish, is much prized for artistic cabinet-work.

It has been known in this country for more than 200 years, and has been largely grown in ornamental collections, being thoroughly hardy, and thriving in most soils and situations. In habit of growth it is sharply conical, the branches abundant, and very dense. Many varieties occur among seedlings, differing chiefly in the colour of the foliage, which varies from the lightest to the darkest green. In exposed situations it assumes a brownish tint in winter.

Var. *glauca* or *argentea*. — This is a beautiful

form, with silvery glaucous foliage, contrasting well with that of the species.

J. drupacea (*the Plum-fruited Juniper*) is indigenous to the north of Syria and mountains of Asia Minor, generally at high elevations, and growing to heights of from 10 to 15 feet. It was introduced into this country in 1820. This very hardy and distinct species has a narrow, conical habit of growth, branched to the ground; the branches are rigid, and clothed with leaves of about ¾ of an inch long, set in regular whorls, glaucous on the upper side, and of a light grassy green on the under.

Any dry, deep soil, not over rich, and moderate shelter from violent winds, will be found sufficient for its wants; and when planted on a lawn as a single specimen, where it has plenty of space, few of our ornamental shrubs are more interesting or attractive.

J. chinensis (*the Chinese Juniper*).—This grand shrub is found very abundantly in high mountain valleys in China and Japan, growing to heights of from 20 to 30 feet, and was first sent to this country in 1820. In this species the male and female forms are very dissimilar, the latter being known as *flagelliformis*, and long after its introduction was regarded as a distinct species. Of the two, the male plant is by far the more ornamental, and undoubtedly one of the handsomest of our hardy coni-

JUNIPERUS.

ferous shrubs, making a magnificent lawn specimen plant, and contrasting most effectively with the dark-coloured shrubs in mixed masses or rows. It has a symmetrical, conical habit of growth, closely branched from the ground upwards; the foliage is generally slightly glaucous, sometimes bright green, and occasionally branchlets of both tints appear on the same plant. It produces even while very young its bright golden catkins in great profusion in May and is then an object of great beauty.

Like most of the other species, *J. chinensis* is of easy cultivation, and will thrive in almost any soil and situation, if the subsoil is sufficiently porous to prevent an undue accumulation of water at the roots.

There are two or three varieties more or less distinct from the species: of these we note the two following, which, though as yet new and not very extensively distributed, will doubtless, when better known and more plentiful, be extensively planted:—

Var. *variegata argentea*, a pretty glaucous form, with a large proportion of the branchlets pure white, the variegation constant, and equally prominent in winter as in summer, while the plant is as hardy as its parent.

Var. *variegata aurea*, sent out two or three years ago by Mr Young of the Godalming Nurseries, is one of the finest variegated Conifers in cultivation, the whole plant suffused with a rich golden hue as bright as that of the Golden Queen Holly. The constancy of its variegation and the hardiness of its constitution remain of course to be thoroughly tested; but there is every reason to believe that it possesses these desirable qualities, and that it will prove a valuable acquisition to the already long list of hardy variegated shrubs.

J. excelsa (*the Tall Juniper*), indigenous to the islands of the Grecian Archipelago, and mountainous districts in several adjoining countries, growing to heights of from 30 to 40 feet. It was introduced about 1806.

In our shrubberies and pleasure-grounds it is a very neat, somewhat slow-growing plant, with a dense, conical habit of growth, the foliage very short, thickly set on the branchlets, and of a peculiar and pleasing glaucous-grey colour. It is a most effective lawn specimen shrub, contrasting well with other shrubs with sombre-tinted foliage, and well worthy of extensive introduction among choice decorative plants.

It is found to succeed best where the soil is deep and rich, and though very hardy, should be planted in a sheltered situation.

Var. *stricta*, a superb variety of recent introduction, so named in reference to the close, tapering, sharply-pointed habit of growth, by which it is distinguished from the parent. It is admirably adapted for geometric gardens, small lawns, or vases, and being quite hardy, will doubtless, when it becomes better known, find a place in most collections.

J. sabina (*the Savin Juniper*).—This well-known species is found wild on the Alps, Apennines, Pyrenees, and other mountain-ranges of Southern Europe, forming a bush of from 7 to 8 feet in height, and has been cultivated in British gardens for nearly 300 years. It is here a broad-spreading shrub, with an irregular outline, the branches numerous, and divided into short branchlets thickly clothed with very minute leaves of the darkest green colour.

Lacking in that formal symmetry of habit so characteristic of many of its allies, and which is so desirable for lawn shrubs, it is nevertheless very ornamental as a variety in the shrubbery, and peculiarly suitable for planting on or in the neighbourhood of rockeries, to which it imparts an interesting and picturesque appearance. It should always have a rich but porous soil, and thrives best in a shady, moderately-sheltered situation.

Of several varieties the best are:—

Var. *variegata*.—This differs only from the parent in having its green branchlets interspersed with others of a pure white colour, and is a most distinct and effective plant.

Var. *prostrata* is a dwarf, trailing form, by some botanists ranked as a distinct species, and found wild in the United States of America: it has a great resemblance to the type in everything except habit of growth. It is a fine plant for rockwork, or for clothing steep banks. The branches are long and very numerous, creeping along the ground, and forming a close carpet never above a few inches thick. It is very hardy, and grows freely when planted in dry and elevated situations.

Var. *tripartita.*—This very distinct sort, said to be of Continental origin, has larger leaves, and a more robust habit of growth, than the species, which it otherwise very much resembles. It is not as yet very widely known, but being of free growth, and remarkably hardy, it will doubtless prove useful for mixed shrubbery borders, particularly in exposed situations, and not less so for clothing dry banks, as it is found to grow, and even thrive, in the poorest soils.

J. recurva (*the Weeping Juniper*), from mountains in Nepal and Bhotan, at elevations of from 8000 to 15,000 feet above the level of the sea, forming a broad densely-branched shrub of about 15 feet high. It was first introduced into our gardens in 1830.

In our pleasure-grounds it is an elegant, feathery bush, with slender branches and numerous branchlets drooping at the points, and clothed with small scale-like leaves, of a deep, sombre green colour. Whether planted by itself on the lawn, or associated with other evergreens in the shrubbery, it is a most beautiful and effective plant, and should be in every collection. Though quite hardy, it requires a sheltered situation, with a shady aspect, and succeeds best in a rich, loamy, but well-drained soil.

Var. *densa*, found naturally in similar localities to the species on the Indian mountains, and introduced only a few years ago, is very different in general appearance, having a dwarfer, more spreading habit of growth, the branches shorter, more rigid, and of a lighter tint of colour. It is a neat little plant for a rockery, and is very useful for small garden-beds.

According to some writers, this plant is the male form of *recurva*, while others assert that it bears fruit in its native habitats, and that it possesses characters sufficiently distinct to entitle it to be classed as an independent species. As it is best known, however, as a variety under this its original name, we retain it as most convenient.

J. squamata (*the Scaly-leaved Juniper*), a native of the Himalayas and Bhotan Alps, at elevations of from 8000 to 15,000 feet, is a decumbent shrub, rarely growing higher than from 3 to 5 feet, first introduced into this country in 1824.

It is thoroughly hardy here even in the most exposed situations, and is one of the prettiest rock-plants in cultivation; the long slender branches are densely covered with short thick leaves, of a slightly glaucous colour. It prefers a dry, sunny situation, and should have ample space for the spread of its branches.

J. tamariscifolia (*the Tamarisk Juniper*), regarded by some authorities as merely a variety of *sabina*, is indigenous to Spain and the mountains of Southern Europe, and was first introduced in 1562.

It has a habit of growth similar to the preceding, rarely rising above 3 feet from the ground, and forming a close, cushion-like bush, with a pleasing silvery-green colour. For banks or rockeries it is invaluable, and so hardy that there are few localities in which it will not succeed. It requires a light, dry soil, and a sunny situation.

J. thurifera (*the Frankincense* or *Spanish Juniper*), indigenous to mountains in Spain and Portugal, growing to heights of from 30 to 40 feet. It was first introduced into Britain in 1752.

This is an exceedingly handsome species, with a close, columnar habit of growth, and abundance of short, twiggy branchlets, clothed with short, rigid leaves, of a light glaucous-green colour. For lawns or prominent positions in shrubberies, few evergreens are more effective.

Though impatient of high winds, it is quite hardy, and grows freely in any locality, and in most soils, if dry and porous.

J. japonica (*the Japan Juniper*), found growing at high elevations on mountains in Japan, from whence it was first sent home about 1860.

As it rarely grows above 3 feet high in its native habitats, and has an erect, compactly-branched habit, with a warm green foliage, it is a very desirable plant for rockwork, or for any arrangement of neat dwarf evergreens. It has stood the test of the past few winters most satisfactorily, and is doubtless as hardy as *chinensis*.

Var. *albo variegata* differs from the species in having its green branchlets intermixed with those of a white or straw colour. It is a pretty little plant, of a compact habit, quite as hardy as its parent, and equally ornamental.

J. rigida (*the Stiff-leaved Juniper*).—This species is indigenous to mountains in Japan and Northern China, where it grows to heights of from 15 to 20 feet. It is quite hardy here, and forms a graceful shrub, of a conical habit of growth; the young shoots are pendulous, and thickly clothed with silvery glaucous foliage. It is undoubtedly a valuable acquisition to our ornamental Conifers, being perfectly hardy, and growing freely in any light, porous soil. The Japanese, by whom it is cultivated in gardens for ornament, call it *moro*, in allusion to its long, drooping branches.

LARIX (THE LARCH).

This genus is composed of several species and varieties of deciduous trees, some of them of large dimensions, and all highly valued for their timber. They are found widely distributed over the colder regions of Europe, Asia, and North America, and are nearly all hardy, and suitable for cultivation in Britain.

Few trees are more ornamental, or more capable of producing grand and striking landscape effects, than the Larches, either when planted in masses, interspersed among other trees, or as single specimens,—their graceful branches and elegant foliage varying in tints from the warm green of early spring to the bright gold of autumn, contrasting admirably with the bolder outlines and more sombre shades of almost every other coniferous tree.

All the known species are handsome enough for the pleasure-grounds, and occupy an important place as ornamental park trees. We confine ourselves, however, to the following, as the most useful for that purpose in our climate :—

L. europæa (*the Common Larch*).—Indigenous to a vast area in Central Europe, inhabiting the mountain-chains from the Alps to the Urals, at elevations of from 2000 to 5000 feet, and growing to heights of from 80 to above 100 feet, with straight, columnar trunks of from 4 to 6 feet in diameter at the base. This well-known and very beautiful species has been cultivated in Britain since 1629, and is of free and rapid growth in almost every variety of soil, provided that it is naturally dry or sufficiently drained to prevent any approach to wetness or accumulation of water at the roots. In cold damp soils it makes little progress, and rarely lives beyond a few years.

With the exception, perhaps, of the Scots Fir, no tree has been more extensively planted in this country for timber than the common Larch, many millions being annually raised from seed both in private establishments and in the various public nurseries for forest planting. In one of the large Scotch nurseries alone, from 1 to 2 tons of seed are required to meet the demands of each spring for young plants. The wood, it is scarcely necessary to say, is of great value, being close-grained and durable, and is extensively used both here and on the Continent in house-building, and for many other purposes where strength and durability are indispensable. Though this species is seldom planted for mere ornament, and would scarcely indeed be placed, unless in a young state, on a moderate-sized, highly-kept lawn, or in a small and select arrangement of the choicer Conifers, it is a magnificent park tree, with a stately picturesqueness of appearance peculiar to itself, and wherever judiciously introduced, imparts to our woodland scenery a richness and grandeur which never fail to arrest the attention of the most careless observer. An old writer remarks: "This is a lofty tree; its branches are slender, and incline downward; the leaves are of a light green; and, like the Cedar of Lebanon, are bunched together like the pencils or little brushes of the painter. In spring, when the leaves and flowers are breaking out, the Larch has a particularly elegant appearance; and in winter it gives variety to a wooded scene by a remarkableness in its naked branches."

Of many varieties which have occurred from time to time among seedlings, differing more or less from the type in habits of growth, foliation, and colour of flowers, we notice the following as the most worthy of attention as an ornamental tree:—

Var. *pendula*.—This very distinct and graceful form originated in this country many years ago from seed saved from the species, and plants grafted upon the common sort are now widely distributed. It has long, pendulous branches, and forms a beautiful weeping tree, well suited for a lawn or any prominent position in a plantation of choice ornamental Conifers.

L. Kæmpferi (*the Golden Larch*), found also in catalogues under the new generic name of *pseudolarix*, is a Chinese species, first sent home in 1852 by Mr Fortune, who thus describes the first specimens he met with in their native habitats: "They were growing in the vicinity of a Buddhist monastery in the western part of the province of Chekiang, at an elevation of 1000 or 1500 feet above the level of the sea. Their stems measured fully 5 feet in circumference 2 feet from the ground, and carried this size with a slight diminution to the height of 50 feet, that being the height of the lower branches. The total height I estimated about 120 or 130 feet. The stems were perfectly straight throughout, the branches symmetrical, slightly inclined to the horizontal form, and having the appearance of something between the Cedar and Larch." This tree is very popular in China for the ornamentation of pleasure-grounds, and is frequently met with in the vicinity of temples. Mr Fortune says: "The Chinese, by their favourite system of dwarfing, contrive to make it, when only 1½ foot or 2 feet high, have all the appearance of an aged Cedar of Lebanon. It is called by them *Kin-le-Sung*, or Golden Pine, probably from the rich yellow appearance which the ripened leaves and cones assume in autumn."

The leaves are about 2 inches long, of a very light green in summer, changing in autumn to a bright golden yellow.

As seen in our collections, it is a very beautiful plant, but so susceptible of injury from early spring frosts that it can only be recommended for the mildest localities. It is, however, worthy of a fair trial, and where it does succeed will amply repay the cultivator for his pains in protecting it during the first few years of its growth, and in all probability it will afterwards ripen its wood sufficiently to withstand any amount of frost it is likely to be subjected to in sheltered situations.

L. microcarpa (*the Small-coned* or *Red American Larch*). — This species, also known as *americana*, is indigenous to a wide area in North America, occurring more or less plentifully from Virginia to Newfoundland, and attaining heights of from 80 to 100 feet. "In the United States it is popularly designated by the name of *Hackmatach*, and the descendants of the Dutch settlers in New Jersey call it *Tamarack*." The timber is close-grained and durable, and in districts where it abounds it is said to be " preferred for general purposes to that of any of the other native Pines." Though quite hardy in most districts, and much planted in parks and pleasure-grounds throughout this country since its first introduction in 1739, it is still only known as an ornamental tree, being of far too slow growth, and too fastidious in regard to soils and situations, for profitable cultivation in woods for its timber. It is, however, very handsome, and distinct enough to warrant its admission to a collection of choice and interesting Conifers. In general appearance it is very similar to the common Larch, but has a more narrowly conical habit of growth, the branches much more slender, and the leaves about half the length of that species.

Var. *pendula*, by some writers regarded as a distinct species, is a very beautiful tree, with foliage larger than that of the common Larch, and with gracefully drooping branches. It is found associated with the species in its native habitats, but is said to occur in greater abundance in the colder regions. Like the species, it is a remarkably beautiful tree, but is not sufficiently at home in our climate to be grown for its timber. Both sorts succeed best in soils rather moist than dry; and, from their tendency to start early into growth, require a north or shady aspect.

LIBOCEDRUS (THE INCENSE CEDAR).

The name of this genus is derived from *libanos*, incense, and *cedrus*, the Cedar, in allusion to the strong odour emitted by the wood while burning. The few species of which it is composed have a general resemblance, both in foliage and habit of growth, to their near allies the *Thujas*, with which they were until recently grouped. They are all remarkably handsome evergreen trees, some of them very lofty, and highly valued for their timber, which is of excellent quality, and extensively used for almost every purpose.

Of the known species, only two can be recommended for outdoor cultivation in Britain: the others, though frequently met with, and deservedly popular as conservatory plants, are much too tender for the rigours of even ordinary winters.

L. chilensis (*the Chilian Libocedrus*), also known as *Thuja chilensis*, is found wild in the high, sheltered valleys of the Andes of Chili, where it forms vast forests, and attains heights of from 60 to 80 feet. The timber is described as being hard and durable, of a fine yellow colour, and pleasantly fragrant. It was first introduced into British gardens in 1848, and has proved one of the most distinct and beautiful of coniferous shrubs; its close, bushy habit, and warm green colour, forming a pleasing contrast to other species of a more diffuse habit and darker hue. It is, however, extremely susceptible of injury from autumn and spring frosts, and in many localities will not survive the winter without protection; though in such as are mild and well sheltered, with the soil sufficiently dry and porous to insure the early ripening of the young shoots, it is found to stand remarkably well, forming a neat, conical, arborvitæ-like shrub, with a great profusion of slender branches, much divided into flat branchlets resembling the fronds of a lycopodium. The leaves are very small, and have a bright shiny-green colour, with a silvery glaucous line along the centre of the under surface. Though, even under the most favourable circumstances, a slow-growing plant in this country—and by no means likely to attain such heights as in its native valleys—it is well worthy of a trial wherever a situation suitable for its wants is at the disposal of the lover of a really handsome and distinct ornamental shrub.

L. decurrens (*the Decurrent-leaved Libocedrus*).— This species, called in some collections *Thuja gigantea*,—a name which really belongs to the plant popularly known as *Thuja Lobbii*,—is indigenous to Upper California, where it is widely distributed, and occurs in considerable abundance at elevations of from 4000 to 5000 feet above the level of the sea. It was first sent home by Jeffrey in 1854, and is one of the best known and most ornamental of that distinguished collector's introductions. In its wild state it is described as a thick bushy tree of from 40 to 50 feet in height, with a stem from 7 to 9 feet in circumference near the ground.

As seen here, it is a plant of great beauty, with a close, columnar habit of growth, the stem thickly clothed with long, flattened, lycopod-like branches, divided into innumerable branchlets of a dark glossy-green colour. Though quite hardy as far as frost is concerned, it is, like many of its allies, impatient of exposure to violent winds, and thrives best in a sheltered situation; and while it grows in almost any kind of soil, it is always found in the highest perfection in deep light loams, with moderately dry and rather porous subsoils.

It is scarcely necessary to add that this is a strikingly effective plant, either in the mixed shrubbery or as a single specimen in the park or lawn; and that its close, spire-like habit of growth, combined with its deep sombre colour, renders it peculiarly adapted for avenue rows, particularly when alternated with other Conifers of a more diffuse form, and of lighter green or glaucous tints, such as *Cedrus deodara, C. atlantica,* or *Cupressus Lambertiana.*

PINUS (THE PINE).

In this important section of the Coniferæ we have a large number of evergreen trees, remarkable not only for the great value of their timber, and the useful secretions which they produce, but for their extreme gracefulness and the grand picturesque effect which they impart to the landscape, either when massed by themselves or blended with other trees of different appearance and habits.

The species are easily distinguished from other Conifers by the great length of their needle-shaped leaves, which, in common with those of the Larch and Cedar, are disposed in sheaths containing a greater or less number, according to the sort: in the case of the Pines they are grouped in bundles of two, three, or five, forming a convenient and sufficiently constant character for reducing the species into three distinct sections.

The flowers are produced in catkins, the male and female being on the same plant, but separate. The cones are for the most part oval in shape, and vary in length from little more than an inch to about a foot.

While the trees are young, and grown singly, they have, with a few exceptions, a handsome conical form, and are regularly furnished with branches from the ground upwards. As they approach maturity, however, the under branches gradually die off, and the trees become broad and bushy at the top. This change is brought about earlier when they are grown close together; so that, if it is desired to maintain the conical form for the longest possible period, careful thinning of the surrounding trees, so as to give the branches the full benefit of light and air, is absolutely necessary.

There are few soils in which Pines will refuse to grow, and even to thrive, if the subsoil is dry, and the land moderately drained; and as the roots rarely penetrate deep into the ground, but run immediately under the surface, they are found to do well in situations where the soil is so shallow that most other coniferous trees would be unable to find sufficient sustenance. As a general rule, however, a good rich soil is the best, and most likely to secure their highest development.

Taking into account the large array of hardy species and varieties now in cultivation, along with the fact that the great majority are highly ornamental and eligible for admission into the choicest collections of decorative trees, it is obviously a difficult matter to make a small selection from the desirable sorts; and though those we are about to recommend are really fine, it is necessary to remark that there are many others which may be grown with great advantage where space and other circumstances admit.

P. austriaca (*the Austrian Pine*) is a noble tree, introduced into Britain in 1835 from the calcareous mountains of Lower Austria. It is also found in Moravia, Transylvania, and more or less abundantly over a wide area in other parts of Southern Europe, attaining heights of from 80 to 120 feet. It is here a dense bushy tree, of free growth in most soils and situations, but is seen in greatest perfection in porous, well-drained loams, and though perfectly hardy, should always be planted where it will be sheltered from the influence of high winds. From its close habit of growth it is invaluable for planting for shelter, or as a nurse to the more tender species. It is one of the few evergreen trees that thrive within the influence of the sea-breeze, and large numbers are now planted in such localities with great success. When grown singly, it spreads out its branches regularly, and becomes in a few years a most symmetrical park specimen tree, invaluable to landscape-gardeners for forming dark backgrounds, or for mixing with other trees of lighter tints of colour. In a paper communicated to the Scottish Arboricultural Society—and published in its Transactions for 1873—by Robert Hutchison, Esq. of Carlowrie, we find the following very judicious remarks upon the importance of the Austrian Pine as an acquisition to our hardy trees: "Its dark, rich green foliage, its dense head of massive contour, its strong side shoots, and its rapid rank growth, all contribute to render it a tree of desirable habit for effective purposes. . . . In localities suitable for its development, and not exposed to heavy winds, *Pinus austriaca* will attain a greater height than the Scots Fir, and is of equally if not more rapid growth. The wood is inclined to coarseness, but is tough and firm in texture, rather knotty, but of more commercial value for country purposes than the timber of equal age of either Larch generally or the Scots Fir." The leaves are usually in pairs, from four to five inches long, and of a very dark glossy-green colour, giving the tree that peculiar sombre shade which suggested the name by which it is in some places popularly known, *Black Austrian Pine*. They are very thickly set on the branches.

Var. *variegata* has its green branchlets intermixed with those of a bright straw colour: it is very effective, and desirable as a lawn specimen.

P. Benthamiana (*Bentham's Pine*).—This magnificent species was discovered by Mr Hartweg growing at high altitudes on mountains in California, and attaining heights varying from 100 to 220 feet, according to soil and situation, the largest specimens being found in dry but rich alluvial soil, and in the shelter of valleys. The timber is reported to be of excellent quality.

It was first introduced into Britain in 1847, and has since then stood the test of our severest winters in almost every district with the greatest success, proving itself perfectly hardy and well adapted for our soils. It should always be planted in a situation sheltered from violent winds, and in light, well-drained ground.

It is a grand ornamental tree, with a broad branching habit of growth, well suited for a spacious lawn or park. The leaves are in threes, from 8 to 10 inches long, of a dark-green colour, and thickly arranged on the branches.

P. Balfouriana (*Dr Balfour's Pine*).—This species was found by Jeffrey on mountains in Northern California, at elevations of from 5000 to 8000 feet, growing in soils composed chiefly of volcanic debris, and attaining heights of from 50 to 80 feet. It was named in compliment to the present distinguished Professor of Botany in the University of Edinburgh.

Though a slow-growing tree, it is perfectly hardy, and beyond doubt one of the most distinct and handsome Pines in cultivation. Its habit of growth is sharply conical, the branches short and very abundant, clothing the stem from the ground upwards. As a single specimen on a lawn, or prominent position in a shrubbery border, it is extremely effective; and but for the scarcity of seedling plants it would be much more frequently met with in collections. Grafted plants, however, are obtainable, and as they form good specimens after a few years, we would recommend cultivators to add it to their collections as a really beautiful and desirable tree. It grows well in light loamy soil with a dry subsoil, and should have protection from the full force of violent winds. The leaves are usually in fives, from $1\frac{1}{2}$ inch to $2\frac{1}{2}$ inches long, thick and rigid, slightly curved inwards, very abundant, and of a light silvery-green hue, giving the plant a peculiarly striking appearance.

P. contorta (*the Twisted-branched Pine*).—This species was first discovered by Douglas, growing abundantly on swampy ground on the north-west coast of North America, particularly at Cape Disappointment and Cape Lookout, forming a small tree rarely higher than about 20 feet. It has since been found near the sea-coast in Northern California, from whence it was introduced to this country, and named *M'Intoshiana*. It is here a slow-growing but very hardy plant, interesting for its long, slender, curiously twisted branches, which give it a picturesque appearance, distinct from that of most of the other species, and renders it valuable as a variety in a collection.

It should be planted in a cool, moist soil, and in a situation rather shady than fully exposed to the sun.

The leaves are usually in twos, about 2 inches long, of a dark-green colour, and very closely set on the branches.

P. Cembra (*Swiss Stone-Pine*).—This very handsome tree is indigenous to the Alps, Eastern Siberia, and several other mountainous regions in Northern Europe, growing at elevations of from 3000 to 5000 feet, and attaining heights of from 50 to 80 feet, according to soils and exposures, the largest specimens being found in sheltered mountain valleys, with deep, rich, alluvial soils.

It was introduced into this country about the middle of the last century, and has proved well adapted for our climate, thriving well in almost every kind of soil, particularly in such as are of a rich loamy character. The timber is soft and easily wrought, but durable and remarkably fine-grained, and is much used in Switzerland and the other countries where it occurs in a wild state, for carving and toy-making. From the high altitudes in which it grows naturally, it has been recommended as worthy of a trial on the hills and more exposed districts in

Scotland, not only for shelter but for its timber. As yet, however, it is rarely planted but as an ornamental tree, and for that purpose it is invaluable, producing beautiful effects when contrasted with the other trees in mixed plantations or as single specimens in the park or lawn. Its habit of growth is sharply conical, the branches short, stiff, very abundant, regularly arranged round the stem, and thickly clothed with leaves. The leaves are in fives, from 2 to 3 inches long, and of a deep green colour.

Var. *pumilo*, sometimes called *pygmæa*, is a curious little plant found naturally in exposed rocky situations on high mountains in Eastern Siberia. It has a close, bushy habit of growth, the branches short, divided into innumerable small branchlets, and very densely clothed with leaves. Being of extremely slow growth—the oldest specimens in its native habitats rarely exceeding 5 feet in height—it is a superb miniature tree or shrub for a rock-garden or small lawn. It is perfectly hardy in the most exposed situations, and though very accommodating as far as soil is concerned, prefers light porous loam, and should always be planted in a sunny aspect.

Var. *siberica*.—This variety is also indigenous to Eastern Siberia, where at altitudes and in bleak exposed situations it occurs in great abundance, and attains heights of from 80 to 100 feet. In habit of growth it closely resembles the species, but is readily distinguished by its much shorter and lighter-green leaves. It is here perfectly hardy, and thrives well in almost every variety of soil, if dry and porous. Though rarely met with in collections, it is a very desirable tree, well worthy of more attention on the part of ornamental planters than it has received.

P. excelsa (*the Lofty Bhotan Pine*).—This magnificent species is indigenous to Nepal, Bhotan, and over a wide range of the Himalayas and other mountains in Northern India, where at elevations of from 5000 to 10,000, or according to some travellers 12,000 feet, it forms vast forests, and rises to heights of from 100 to 150 feet, dwarfing down at the extreme altitudes, however, to a mere bush.

It was first sent to this country in 1823, and has since been extensively planted in parks and other ornamental grounds, proving itself perfectly hardy in almost every district, and of free and even luxuriant growth in almost every variety of soil, if light, deep, and well drained.

In all stages of its growth it is a singularly graceful plant, with a broadly conical habit of growth, the leading shoot well relieved from the top, the branches very abundant, and arranged in regular whorls round the stem from the ground upwards. The branchlets droop slightly at the extremities, and being thickly clothed with slender leaves of a shining silvery-white colour, from 5 to 6 inches long, the tree has a soft, fleecy appearance, very distinct from that of any of the other species. Planted as a single specimen on a spacious lawn, it has a grand effect, and it cannot be too highly recommended for even the smallest collection of coniferous trees.

Var. *rigida* is a distinct form of the species, said to occur at very high altitudes in Nepal: the leaves are much shorter, and the habit of growth less diffuse than the type. It is as yet a scarce and little known plant.

P. Fremontiana (*Colonel Fremont's Nut-Pine*) is a singularly interesting species, introduced in 1848 from California, where it occurs in great abundance in several of the high mountain-ranges, at elevations of from 4000 to 7000 feet, rarely attaining a greater height than 20 feet. The seeds, which are of an agreeable flavour and very nutritious, form the principal winter and spring sustenance of the Indians who inhabit the mountains. In this country it has proved perfectly hardy, thriving best in light dry soils and open situations, forming a neat, distinct-looking, conical tree, densely feathered with branches. It is, however, a very slow grower, and should only be planted on the margins of walks, or where there is no probability of its being hidden or overshadowed.

by the more robust species. The leaves are generally in threes, sometimes in fours or fives, and not unfrequently in young plants singly, from 1½ inch to 3 inches in length, very stiff, and glaucous green.

P. insignis (*the Remarkable Pine*), discovered and sent home by Douglas in 1833 from California, where he found it in several districts, but chiefly near Monterey, on mountains near the coast, and rising to heights of from 60 to 100 feet. Its merits as a distinct and handsome ornamental Pine can scarcely be overrated, forming as it does a fine symmetrical, though by no means formal, tree, densely furnished with branches; while the warm grassy-green foliage with which it is so abundantly clothed contrasts pleasingly with the more sombre hues of most of the other species. From its natural tendency, particularly while young, to make late autumnal growths, and as a consequence to be damaged by frost, some planters have been deterred from introducing it extensively, or from giving it that prominence which they would otherwise do; hence its absence in many of the best collections of Conifers. It is, however, a very hardy species, and needs but to be planted in such a situation as will secure its growths being thoroughly ripened before winter. A high, airy, yet sheltered position, as far as practicable shaded from the morning sun, and a dry, porous subsoil, will suit it admirably, and render it proof against any amount of frost it is likely to be exposed to in this country. In one of a series of admirable papers contributed to the 'Gardeners' Chronicle' in 1872, Mr Fowler, head-gardener to the Earl of Stair, Castle Kennedy, thus speaks of this Pine: "This gorgeous tree is ever pleasing, whether seen in midwinter, draped with its beautiful grass-green foliage, decked by its clusters of cones of varied hues, or in midsummer, when its darker old and lighter-coloured delicate young foliage contrast so agreeably. This is a Pine which would prove hardy in many localities where it is not supposed to be capable of standing, if a little more attention was bestowed on it for a time after being planted. It appears to increase in hardiness after being well established, as not a few other Conifers are well known to do. I have frequently observed young plants, a few feet in height, suffer from, and in some cases be altogether killed by, cold frosty winds, when larger and stronger plants were altogether unscathed." He adds that some specimens planted about eighteen years ago at Castle Kennedy are now upwards of 36 feet in height, measuring at 2 feet from the ground 5 feet 6 inches round the bole, and that these are as little injured by the cold cutting breezes as is the *P. austriaca* or the *P. laricio*. The leaves are from

4 to 6 inches long, and generally, though not always, grouped in threes.

P. Jeffreyii (*Mr Jeffrey's Pine*), found by Jeffrey in the Shasta Valley, Northern California, and sent home by him in 1848, is a noble tree attaining the height of 150 feet. It is quite hardy here, and forms a remarkably distinct-looking specimen tree, admirably adapted for parks and extensive lawns. In general appearance it resembles *Benthamiana*, but is very different from that species, having more slender branchlets, and the foliage of a deep bluish green. It will grow in almost any soil if well drained, but prefers a light sandy loam, with a porous subsoil.

The leaves are usually in threes, from 8 to 9 inches long, and densely set on the points of the shoots.

P. koriensis (*the Corean Pine*) is a small, slow-growing species, never exceeding 20 or 30 feet in height, sent to this country from Japan, where, as well as in China, it is extensively cultivated in gardens. Its natural habitat is mountains on the peninsula of Corea, on the sea-coast, where it is found in immense forests, producing abundance of seeds, which are an important and wholesome article of food to the natives.

Resembling *strobus* in its slender, thread-like

leaves, and *cembra* in its compact, conical habit of growth, it is one of the most beautiful of our hardy Pines, and may be introduced into the most prominent positions on lawns, and in the highest-kept pleasure-grounds, with the happiest effect.

It prefers a deep, loamy soil, moderately dry; and though hardy enough to withstand our severest frosts, should be planted in a situation sheltered from strong winds.

The leaves are usually in fives, densely clustered at the points of the branchlets, and of a light, glaucous-green colour.

P. Lambertiana (*Lambert's Pine*).—This is a gigantic tree, widely distributed over Northern California and North-West America, attaining in some districts heights of from 150 to above 200 feet, with trunks of from 20 to 60 feet in circumference. It was first discovered by Douglas, near the source of the Multnomah river, in 1827. In his description of the trees, as he saw them in their native habitats, he says: "One specimen which had been blown down by the wind, and which was certainly not the largest, was of the following dimensions: its entire length was 215 feet; its circumference at 3 feet from the ground was 57 feet 9 inches, and at 134 feet from the ground 17 feet 5 inches. The trunk is unusually straight, and destitute of branches about two-thirds of its height. The bark is unusually smooth for such large timber, of a light-brown colour on the south, and bleached on the north side." The wood is reported as soft, light, and easily worked. The resin, which it produces in great quantity, has a sweet taste when roasted, and makes an excellent substitute for sugar. The seeds are large, very wholesome, and, cooked in a variety of ways, form an important article of food to the natives of the regions where the tree abounds.

In our parks and pleasure-grounds it has, particularly while in a small state, a general appearance suggestive of the Weymouth Pine, but is easily distinguished from that species by its longer, more decidedly glaucous leaves, as well as by its more robust branches, and more broadly conical habit of growth. For planting in parks or spacious lawns, either as a single specimen or blending with other species in masses for the creation of landscape effect, it has few rivals, its peculiarly beautiful light-green colour, graceful feathery branchlets, and symmetrical outline, rendering it always a pleasing and effective feature in any ornamental arrangement. It should be planted in light, well-drained soil; and though quite hardy, and of easy culture in most districts of Britain, it thrives best in moderately-sheltered situations. The leaves are usually in fives, and vary in length from $3\frac{1}{2}$ to 5 inches.

P. laricio (*the Corsican Pine*) is an erect, lofty

tree, introduced from the island of Corsica in 1814, and also indigenous to mountains in Calabria, and over a vast range in Southern Europe, forming magnificent forests at elevations of from 4000 to about 6000 feet, and growing to heights of from 80 to 150 feet. The wood is reported as pliant, easily worked, very resinous, and durable, and is much valued on the Continent for house-building purposes.

In this country it is as hardy as the Scots Fir, quite as accommodating in regard to soil, of much more rapid growth, and less liable than any other of the Pine tribe to the attacks of hares and rabbits. During the past few years it has been growing in favour as a forest-tree; and as young plants can now be procured nearly as cheap as Scots Fir or Larch, it is already extensively planted. In the course of an eminently practical paper which appeared in the Transactions of the Scottish Arboricultural Society for 1868, Mr John M'Laren, forester to the Earl of Hopetoun, thus states his experience of it in Linlithgowshire: "I planted a considerable number of it on a piece of ground called from its bare aspect 'the Heather Muir,' and which a few years ago was covered with heath. The soil, as may well be conceived, was not of the best description for trees. Its surface was composed of that thin, light, sandy soil common to heath tracts, and having a subsoil of stiff, hard clay. Along with these were planted Spruce, Scotch Fir, and Larch; and although many of these were injured by hares and rabbits, with which the place abounds, none of the Corsican Pine were injured. . . . They show, too, every symptom of vigorous and sustained growth, are outstripping the others in size, are cleaner in the stem than the Scotch Fir, are altogether free of the insect commonly called aphis, have none of those ulcerated wounds so common to the Larch, and even during this trying season many of them grew upwards of 18 inches, while their neighbours—the Scotch Fir, &c.—have made on an average only 14 inches. . . . From its more rapid growth, and larger dimensions when full grown, it will produce more timber in agiven period than the Scotch Fir; and therefore, even when viewed merely in the light of an economical question, it is well worthy of general cultivation."

Ever since its introduction, it has been much valued as an ornamental tree, and fine specimens are frequently found in parks and other pleasure-grounds. From the rapidity of its growth, however, it is only desirable as a lawn specimen while in a young state. The habit of growth is straight and conical, the stem furnished with numerous short spreading branches, disposed in regular whorls from the ground upwards; and when allowed plenty of space, forms a remarkably beautiful object. Though very accommodating in regard to soils, it prefers such as are light and porous: a rich sandy loam with a cool subsoil suits it admirably, and in every case the ground should be sufficiently drained to prevent water stagnating at the roots. The leaves are in pairs, from 4 to 6 inches long, curiously twisted, and of a bright green colour.

Var. *calabrica*.—This form is found in similar situations to the species on high mountains in Calabria. It differs chiefly in having longer leaves, which are more abundantly set on the branches. It is equally hardy, quite as desirable as an ornamental tree, and if planted in sufficient quantity, would doubtless be as important as a producer of good workable timber.

P. muricata (*the Bishops' Pine*).—This very distinct Pine, introduced from Upper California in 1848, was first discovered by Dr Coulter at San Luis Obispo, at an elevation of 3000 feet above the level of the sea, and distant about 10 miles from the shore. It was subsequently found by Jeffrey on the Siskiyou Mountains, at an elevation of 7500 feet, growing in damp soil.

It is perfectly hardy in this country, and thrives well in any rich deep soil, forming a very ornamental, densely-branched, conical tree, of somewhat slow growth. The leaves are usually in pairs, not very thickly set on the branches, from 3 to 4 inches in length, stiff, rather broad, and of a deep green colour.

P. macrocarpa (*the Large-coned Pine*), indigenous to California, and also known as *Coulterii*, is a magnificent tree, noted for the large size of its cones, which are sometimes 14 inches in length, 6 inches in diameter, and weighing from 3 to 4 lb. Nuttall says that "it was first discovered by Dr Coulter on the mountains of Santa Lucia, near the mission of San Antonio, in 36° of latitude, within sight of the sea, and at an elevation of between 3000 and 4000 feet above its level. It was accompanied by the *P. Lambertiana.*" It is also found plentifully in various other parts of California, on high mountains near the coast, forming a lofty tree of from 80 to 100 feet in height, with a trunk of from 3 to 4 feet in diameter near the ground. Since its introduction in 1833, it has been widely planted in almost every district of the country, and has been found not only perfectly hardy, but of free growth in most soils and situations, if dry and moderately sheltered from the full force of the wind. It is a most distinct and striking landscape-tree, as well as a handsome specimen for a lawn. The branches are in regular whorls, horizontal, but slightly turned up at the extremities; and though not very abundant, the dense clothing of long leaves gives the plant a graceful and furnished appearance. The leaves are usually in threes, from 8 to 10 inches long, rather rigid, and of a bluish-grey colour.

P. monticolo (*the Mountain Pine*), introduced from Northern California in 1831, where it grows on the highest mountains in poor, shallow soil, composed of the debris of granite rock, and rising to heights of from 60 to 100 feet, with a circumference of from 4 to 6 feet near the ground. It is very hardy here, and forms a compact, conical tree, somewhat resembling the Weymouth Pine, but far denser in habit, and with shorter and lighter-coloured leaves. The branches are disposed in regular whorls, very abundant, and well clothed with foliage. Few of the species are more beautiful, or more deserving of being extensively used for the enriching of either woodland or garden scenery. It grows vigorously in most soils, if loose in texture and the subsoil porous. The leaves are usually in fives, from 3 to 4 inches long, slender, and of a distinct silvery-glaucous colour, especially while young.

P. ponderosa (*the Heavy-wooded Pine*), sometimes called *Beardsleyi*, is a noble tree, found in great abundance on the north-west coast of North America, growing to heights of from 80 to 140 feet. It was first sent home in 1826. The timber is reported as very durable, and so heavy as to sink in water. It is thoroughly hardy here, and grows freely if planted in rich loamy or alluvial soils, and in situations not too much exposed to wind. The branches are remarkably robust, rather drooping, regularly disposed in whorls, and not very numerous. As an ornamental tree it is rather picturesque than beautiful, is singularly effective as a single specimen, and forms a fine contrast when associated with other species of a denser habit of growth. The leaves are usually in threes, from 8 to 10 inches long, broad, slightly twisted, and bright green.

P. pinaster (*the Cluster-coned Pine*), also known as *maritimus*, is a handsome tree of from 50 to 80 feet in height, found abundantly all along the northern borders of the Mediterranean Sea, at elevations ranging from the shore to nearly 3000 feet above its level. The specific name *pinaster* (Star Pine) refers to the peculiar arrangement of the cones in whorls round the branchlets. It is largely cultivated in France and several other countries in Southern Europe, on sandy plains near the sea; and its timber, though not very durable, is used for a great variety of common purposes.

Since its introduction to this country nearly 300 years ago, it has been extensively planted, growing freely in almost every district where the soil is of a dry sandy or gravelly character, forming a beautiful bushy tree, admirably adapted for seaside planting, both for ornament and shelter. In damp, peaty, or cold clay soils it makes little progress, and becomes little more than a dwarf stunted shrub. In favourable circumstances it produces its cones freely, which, being large and prominent, give the tree a pretty appearance.

The leaves are usually in pairs, from 6 to 8 inches long, and of a warm grassy-green colour.

Of a number of named varieties the following is the most useful: it is quite as hardy as the species, very distinct, and thrives under similar circumstances:—

Var. *Hamiltonia*, at first supposed to be a distinct species, was introduced by Lord Aberdeen in 1825, is a remarkably beautiful tree, differing from the type in its leaves being of a paler green, and in its cones being shorter and more ovate.

P. radiata (*the Radiated-cone Pine*), discovered by the late Dr Coulter in 1830 growing on mountains near the sea in Upper California, and attaining the height of 100 feet. In this country it has proved a very hardy and highly ornamental tree, well furnished to the ground with branches. In general appearance it resembles *insignis*, of which it was formerly regarded as a variety. It is, however, quite distinct, the leaves being much shorter, and the cones nearly three times the size of that species. It has, moreover, been found to be much hardier and less susceptible to damage from early autumn frosts, thriving well in more exposed situations, and even with moderate shelter within the influence of the sea-breeze. It should be planted in a dry sandy soil. The leaves are usually in threes, from 3 to 5 inches long, slightly twisted, densely set on the branches, and of a rich green colour.

P. rigida (*the Stiff-leaved Pine*), introduced in 1750 from America, where it is found in great abundance from New England to Virginia, growing in almost every kind of soil and altitude, rising to heights of from 50 to 80 feet. It is quite hardy here, and has a very neat habit of growth, well furnished with robust branches, covered densely with foliage. It is a fine plant for growing as a single specimen, thriving well in most soils if well drained. The leaves are usually in threes, very stiff and sharp-pointed, from 3 to 5 inches long, and dark green.

P. Sabiniana (*Mr Sabine's Pine*).—This beautiful tree was first discovered by Douglas on the Cordilleras of California, growing at high elevations, and attaining heights of from 80 to 150 feet, the finest specimens being in moist soils. It is also found abundantly in several districts in Upper California, and not unfrequently close to the sea. Since its introduction in 1832 it has been widely distributed, and has proved quite hardy all over the country, growing freely when planted in deep, rich, loamy soil, and in situations airy but sheltered from cutting winds. The branches are somewhat slender, but numerous, and regularly set round the stem. The leaves, which are usually in threes, are from 8 to 10 inches long, twisted, and of a pleasing glaucous-grey colour. They are very dense at the points of the branches, a peculiarity which allows the fine violet bloom with which the bark is covered to be seen to advantage. It is a handsome plant in all stages of its growth, and indispensable where scenic effect is desired, either in the Pinetum or in mixed ornamental grounds.

P. strobus (*the Weymouth Pine*), a tall tree, found abundantly on mountains in Canada and the United States. It attains heights of from 100 to 150 feet, and produces the timber known as American white Pine: it was introduced into this country in 1705. It is perfectly hardy here, and forms a straight, handsome tree, with short branches regularly arranged round the stem from the ground when young, the lower ones gradually dying off as it increases in height, and the old specimens assuming a broad, bushy-headed, rather than a conical shape. The bark is smooth, and of a light greenish-grey colour; the leaves are in fives, from 3 to 4 inches long, very slender, and glaucous green. In parks and ornamental plantations few trees are more effective; and while it cannot be recommended for cold damp clays or peaty soils, it is of remarkably free growth in such as are dry and porous.

Var. *nana* is a miniature of the species, rarely found above 2 or 3 feet high, with a flat, tabular head, and is an interesting little bush for a rock-garden or small lawn. It is quite hardy, and, like its parent, requires a dry, sandy, or gravelly soil.

P. sylvestris (*the Scots Pine*).—This well-known and useful tree has a wide geographical range on the continent of Europe, and is the only true Pine indigenous to Britain, being found wild on the mountains of Scotland, varying in height, according to the soil and situation, from a mere bush to a lofty tree of from 80 to 100 feet, with a trunk of from 2 to 4 feet in diameter near the base. The wood is close-grained, strong, and very durable, and extensively used all over Europe for house-building purposes. Though from its extensive cultivation—not only in Britain, but all over the Continent—for timber, it is not usually classed among ornamental trees, it should never be altogether omitted from such collections. None of the species has a finer effect on the landscape, either when planted in groups or as single specimens—its dark foliage and bold irregular outlines commending it to all who have a taste for the picturesque in scenery; while its extreme hardiness and dense foliation render it invaluable as a shelter tree, or as a nurse to the more tender sorts.

There are many very distinct and desirable

varieties in cultivation, differing more or less from the species in colour of foliage and habits of growth : among these may be noted :—

Var. *aurea*, with a beautiful golden variegation, is a pretty plant for a small lawn, being of slower growth than, but equally handsome in habit as the species.

Var. *nana*, a curious dwarf form, with a globular habit, rarely growing higher than from 2 to 3 feet : it is only cultivated as a curiosity, and is usually met with in rock-work collections.

Var. *pendula* differs from the parent in its branches being more slender, giving the tree a drooping appearance.

Var. *horizontalis* has its branches disposed in a horizontal direction from the stem. This is the Speyside and Braemar variety, and regarded as the hardiest, and as producing the finest timber.

Both the species and varieties accommodate themselves freely to almost any kind of soil and situation, but prefer a cold, stiffish clay if well drained, and an elevated though not too much exposed situation. The leaves are usually in pairs, from $1\frac{1}{2}$ inch to $2\frac{1}{2}$ inches long, rigid, and vary according to the variety from a dark-green to a slightly glaucous colour.

P. tuberculata (*the Tuberculated-coned Pine*), from Upper California, first discovered and described by

Dr Coulter some forty years ago, and more recently by Jeffrey, who sent home seeds in 1848. It is found in a great variety of situations, very frequently on the sea-beach, from near its level to elevations of from 3000 to 5000 feet. It rarely rises higher than about 30 feet, with a trunk of about a foot in diameter. Though quite hardy enough for our climate, it is of very slow growth, but forms a nice, distinct-looking, lawn specimen tree, of a conical shape, and well furnished with branches disposed irregularly over the stem. The leaves, which are densely set on the branches, are usually in threes, about 4 inches long, stiff, broad, and of a bright shiny-green colour.

PICEA (THE SILVER FIR).

This interesting section of the Coniferæ contains some of the most beautiful of our hardy evergreen trees. In general appearance and habits of growth they resemble the Spruce Firs, with which they are by some writers associated. They have, however, several well-marked and constant characters, which seem to justify their separation into a distinct genus, and which, at least, form a convenient guide to their popular identification. Their leaves are for the most part longer and more flattened, silvery beneath, and disposed more obviously in two rows than those of the Spruces; while the cones are invariably erect, produced on the upper side of the branches, and deciduous, the scales falling off when they reach maturity.

Like the Spruce Firs, they have a wide geographical range in Europe, Asia, and America, but are found in more temperate regions, or at least in more sheltered situations. A large proportion of the species are quite hardy, and do well in most districts in this country; and though, with the exception of the common Silver Fir (*P. pectinata*), as yet of little repute as timber-trees, they are all highly appreciated and extensively planted for decorative purposes,—making beautiful lawn specimens, and producing the finest effects, either as grouped by themselves or blended with other Firs or deciduous trees in mixed plantations.

With few exceptions, they thrive best and form the finest trees in low-lying, sheltered places, when the land is moderately moist without being wet, and of a deep, rich, loamy or alluvial character. In thin, dry, gravelly or sandy soils, and on high situations, where many of the Pines will succeed well, the Silver Firs become mere stunted bushes, and rarely live longer than a few years.

In the earlier stages of their growth, and for a considerable time after transplanting, they make very slow progress; but after attaining the height of a few feet, and the roots becoming thoroughly established in good soil, other circumstances being favourable, they shoot up with great rapidity. Several of the species, though otherwise quite hardy, have a tendency—particularly while young—to start early into growth, and, as a consequence, to suffer from late spring frosts. These should always be planted in a situation where they will be shaded from the early morning sun, the influence of which, if brought to bear directly upon the young growths while still frozen, and before the temperature rises so as to thaw them gradually, is more fatal than the frost itself.

Many of the hardiest and most ornamental of the species are from Northern California, and though some of these have been known in Britain for more than forty years, it is a remarkable fact that, with the exception of such as are bearing cones, they are still among the scarcest of the Fir tribe. Seeds have again and again been imported, and within the last few years in considerable quantity; but the number of plants raised has been comparatively limited. In the case of *P. amabilis*, it is even questionable if a single seed has germinated since the few first sent home by Douglas in 1831. This arises not from any difficulty experienced by collectors in obtaining cones, as they can be had in abundance, but from the attacks of a small insect which deposits its eggs in the seeds before they are ripe, and while the cone is yet soft and green. The maggot is hatched as the seeds become matured, and feeds upon the kernel; and though apparently perfect and sound before their despatch to this country, they are found on their arrival, in many instances, to be so perforated and hollow as to render germination impossible.

The following are among the hardiest and most desirable of the species:—

P. amabilis (*the Lovely Silver Fir*).—This, one of the grandest of the Californian Firs, was first discovered by Douglas on the mountains near Fraser's River, in Northern California. Jeffrey afterwards found it at elevations of from 3000 to 4000 feet, rising to heights of from 150 to 250 feet—with trunks from 5 to 7 feet in diameter near the ground —the straight, arrow-like stems in many instances clear of branches for 100 feet. It is quite hardy here, growing freely in rich deep soils, if not too dry; and from the few specimens which we have had the opportunity of seeing, we are convinced that it could not be named more appropriately. It is indeed a "lovely" tree, with a slender, conical habit, thickly branched, and profusely covered with foliage. The leaves are about an inch long, incurved on the upper sides of the shoots, of a dark glossy-green colour on the upper side and silvery beneath.

From the cause we have already stated, it is, so

far as seedling plants are concerned, as yet extremely rare, and planters have to content themselves with young plants raised by cuttings or grafted upon some of the other species, which, though requiring a long time to form leading shoots, become ultimately handsome plants, and repay all the patience and care that have been bestowed upon them.

P. balsamea (*the Balm of Gilead Fir*).—This well-known and handsome species is indigenous to Canada, Nova Scotia, and the northern States of America, and was first introduced to this country about 1696. It constitutes vast forests in low sheltered valleys and by the sides of rivers, growing in rich, alluvial, dampish soil, and rarely rising higher than from 30 to 50 feet. It is quite hardy here, and though little valued as a forest or timber tree, it gives a pleasing variety to mixed plantations, and forms, particularly while young, a remarkably handsome lawn specimen, having a neat conical habit, somewhat slender, but well furnished with bushy branches to the ground, and profusely clothed with short, bright green leaves, which are silvery on the under side. Having the unfortunate tendency of making early growths, it should always be planted in cool, stiffish soil, and in a north or west aspect, where it will bear any amount of frost, and prove all that could be desired as a pretty ornamental tree.

P. cephalonica (*the Mount Enos Fir*).—This fine species was first sent to Britain from Mount Enos, in Cephalonia, in 1824, where, at an elevation of 5000 feet above the level of the sea, it grows to the height of about 60 feet, with a trunk of from 3 to 4 feet in diameter. It is also found in considerable abundance in high mountainous districts in Greece. It is quite hardy here, though, if planted in unsuitable situations, liable to damage from late spring frosts. To insure success, it should have a shady exposure and a stiffish soil. It is a distinct and beautiful tree, admirably adapted for growing as a single specimen, having a broad conical habit, branched to the ground, and most profusely covered with foliage. The leaves are very short, stiff, and sharply pointed, dark green above, and silvery beneath.

P. cilicia (*the Cilician Silver Fir*).—This beautiful tree, though regarded by some writers as only a variety of the common Silver Fir, to which it bears some resemblance, is nevertheless so distinct, and forms such an attractive ornamental tree, as to be worthy of introduction into the most limited collections. It is found on Mount Taurus, and in many of the other mountains in Asia Minor, growing in high but sheltered situations, and in rich loamy soil. It rarely attains a greater height than about 50 feet. It is quite hardy in this country if protected from violent winds, and makes a neat lawn specimen, of a close, bushy, conical habit of growth, regularly branched to the ground. The leaves are of a darkgreen colour above and silvery beneath.

P. Fraseri (*Fraser's Silver Fir*).—This elegant species is indigenous to high mountains in Virginia, Pennsylvania, and New England, forming a small conical tree of from 15 to 20 feet in height. It was first sent to this country in 1811.

Though closely resembling *balsamea*, of which it was at first regarded as a variety, it is very distinct, having much smaller cones; shorter, darker green, and less glaucous leaves.

It has a tendency to grow early, and consequently to suffer damage from our spring frosts; but when planted in cool soils, and in north aspects, it generally thrives well, and forms a handsome, conical, dwarf tree of slow growth, and very desirable as a lawn specimen.

Var. *hudsonica*.—This is a peculiarly interesting dwarf form, rarely found even in its native mountains higher than from 2 to 4 feet. It is indigenous to high elevations in the Hudson Bay Company's territory, from whence it was first sent home about 1820. In this country it is a broad bushy shrub, very densely branched with small flat leaves, dark green above and slightly glaucous below. It is a fine rock-work or small lawn plant, and ought to be better known.

P. grandis (*the Great Silver Fir*), first sent home by Douglas from Northern California in 1831, is a majestic tree, found growing in moist valleys and banks of rivers to heights of from 200 to 280 feet, with a diameter near the ground of 5 feet. It is one of the hardiest of our exotic trees, growing with great luxuriance in any situation where it is moderately sheltered, and the soil deep and rich. Few of its tribe form more beautiful single specimens, or are more likely to produce, as they increase in size, a more striking landscape effect. It has a fine conical habit of growth, densely branched to the ground. The leaves, which thickly cover the shoots, are from 1 to $1\frac{1}{2}$ inch long, flat, deep shiny green above and silvery beneath. Seedling plants of this interesting tree are as yet rarely to be had.

P. lasiocarpa (*the Woolly-scaled Silver Fir*), introduced as recently as 1860 from Northern California, where it grows to heights of from 200 to 250 feet. It is quite hardy in every district in Britain, and is one of the most beautiful of the tribe, forming a symmetrical, conical specimen, regularly and thickly branched to the ground. The leaves are somewhat sparsely distributed over the branches, from $1\frac{1}{2}$ inch to 2 inches long, broad, and of a distinct light-green colour, contrasting well with the other species, and along with its handsome form rendering the plant a conspicuous and pleasing object, whether met with on the lawn or mixed plantation. It is only necessary to add that it is indispensable in every collection, and that it accommodates itself to almost every variety of soil, and is by no means fastidious in regard to situation.

P. magnifica (*the Superb Silver Fir*), known also as *robusta magnifica* and *nobilis robusta*, is another of those grand introductions from Northern California within the last few years, where it is described as a stately tree, rising sometimes to a height of 250 feet. When first sent home, it was regarded as only a distinct variety of *nobilis*, which when in a young state it resembles very much; and it is still a disputed point whether it has a right to be ranked as an independent species. It is, however, a remarkably handsome and very popular tree, well worthy of cultivation in every collection. It has proved itself perfectly hardy, and will thrive and grow with the greatest luxuriance in ordinary rich soils, while, from its lateness in starting into growth, it may be planted in open sunny situations. In common with *nobilis* and others of its class, it has a symmetrical, conical habit of growth, the leading shoot well relieved from the branches; the leaves are light green, longer, broader, and not so abundant as on that species.

P. nobilis (*the Noble Silver Fir*), found wild on the north-west coast of North America and in Northern California, growing on mountains at elevations of from 5000 to 8000 feet. It was introduced into this country in 1831.

In its native woods it rises to heights of from 150 to 200 feet, with a diameter near the ground of from 3 to 4 feet. Douglas, by whom it was first discovered, says of it: " This singular species is a majestic tree, forming vast forests on the mountains of Northern California, and produces timber of an excellent quality. . . . I spent three weeks in a forest composed of this tree, and day by day could not cease to admire it."

It is here as hardy as the common Spruce or Scots Fir, growing freely in almost all soils if deep and moderately rich, and none of the ornamental Conifers are more frequently met with in our parks and pleasure-grounds. A distinguished arboriculturist thus speaks of its value as a hardy tree: "Its magnificent appearance, rapidity of growth, as well as hardiness when planted at different elevations and exposures, all go to prove the truth of the high encomiums which have been bestowed upon it; and we consider it a tree well worthy of being introduced to a greater extent than it has ever yet attained, not only on account of its great beauty, but also with the view of its ultimately becoming a valuable acquisition to our forest-trees." The branches are regularly arranged round the stem, spreading out horizontally and very numerous, each tier forming a level platform round the tree. The leaves, with which every shoot is covered so thickly as to have a crowded appearance, are from 1 to $1\frac{3}{4}$ inch long, of a peculiar bluish-green tint on the upper surface, and silvery on the under. As a lawn specimen it is unrivalled, whether as regards symmetry of form or colour of foliage; and when associated with evergreen trees of lighter shades in avenue rows, it produces strikingly interesting and pleasing effects. There are few lovers of handsome trees who will question the appropriateness of the name which Douglas conferred upon it, as, beyond doubt, it is one of the noblest and most beautiful of our hardy trees.

P. Nordmanniana (*Nordmann's Silver Fir*).—This grand species was first discovered by Professor Nordmann on the Adshar Mountains, where it occurs in great abundance. It is also found in great forests on the Crimean mountains, and on those to the east of the Black Sea, at elevations of from 3000 to 6000 feet above its level, attaining heights of from 80 to 100 feet. The timber, being hard and durable, is much used in house-building. It was first sent to this country in 1845, and having been widely distributed and largely planted in almost every district, its perfect hardiness and adaptability to our soils are now thoroughly demonstrated, fully justifying the high estimate formed of it by one of our highest authorities on all matters connected with arboriculture, who a few years after its introduction remarked: "It starts into growth later than the common Silver Fir, and is consequently less liable to suffer from spring frosts. Its timber is of better quality; and as there is an increasing facility for obtaining seeds from its native forests, we have no doubt whatever that it will soon take a prominent position among our most profitable timber-trees, and be planted by the million in woods."

As an ornamental tree it is exceedingly beautiful, having a broad, conical habit of growth, the branches very abundant, spreading out horizontally, and regularly arranged round the stem from the ground upwards. The leaves are about 1 inch long, flat, of a bright shiny-green tint above, and glaucous beneath. It is a grand object standing alone on a lawn, or any other situation where, free from contact with other trees, it has sufficient space to develop its branches on every side. It prefers a cool, stiff, loamy soil, and is always seen in greatest perfection when sheltered from the full force of the wind.

P. pectinata (*the Common Silver Fir*).—A well-known and beautiful tree, introduced in 1600 from the Alps. It is also found in vast forests on the

Apennines, and other mountain-ranges in Central and Northern Europe, at elevations of from 2000 to 5000 feet, and growing to heights of from 50 to 150 feet. Its timber is soft, and not very durable, but is extensively used for a variety of common purposes. It has long been extensively planted in this country as a forest-tree, and in some districts succeeds well, the finest specimens being found in sheltered valleys, where the soil is of a rich alluvial character. It is very impatient of the sea-breeze, and should never be planted on dry sandy or rocky land.

Under favourable circumstances it forms a handsome tree, and though common in our woods, is no less worthy than any of its congeners of a place in the choicest ornamental collections. When grown singly it assumes a neat conical form, regularly and densely branched from the ground upwards. The leaves are about 1 inch long, of a shiny dark green above, and silvery beneath. As it commences to grow early, and frequently suffers from spring frosts, it should always be planted in a shady situation.

P. pichta (*the Pitch or Siberian Silver Fir*) is a small tree, seldom exceeding 40 feet in height, found wild on the mountains of Siberia and the Altai, in immense forests, at elevations of from 2000 to 5000 feet. It was introduced in 1820. It is here a distinct and pretty ornamental tree of conical shape, abundantly branched to the ground. The leaves are thickly set on the shoots, dark green above, and lighter but not silvery beneath. In low-lying places it starts early into growth; but if planted on high, rather exposed situations, and in stiffish soil, it will succeed well, and form a fine vigorous specimen.

P. pinsapo (*the Pinsapo Silver Fir*).—Found abundantly on the higher mountains of Spain, principally on northern exposures, and sometimes reaching to their summits, where the snow lies for five months in the year, and growing to heights of from 40 to 70 feet. It was introduced into this country in 1838, and has been found to be perfectly hardy in our severest seasons. Few trees are more ornamental and effective when planted singly, or sufficiently apart from other trees that all the branches have plenty of room to develop themselves, forming as it does a symmetrical conical tree of very dense habit, with the branches divided into innumerable ramifications from the ground upwards, so as to hide the main stem, giving it a remarkably distinct and interesting appearance. The leaves are very short, rarely more than 1 inch, cylindrical in shape, very sharp-pointed, regularly and thickly disposed round the branches, and of a pleasing light-green colour. It is of slow growth, but thrives well in any ordinary rich soil, and prefers an open situation if moderately sheltered.

P. pindrow (*the Upright Indian Silver Fir*).—This

beautiful species was introduced from the Bhotan Alps, where it is said to occur in great abundance at elevations ranging from 1000 to 12,000 feet above the level of the sea, and in sheltered valleys rising to heights of from 80 to above 100 feet.

In favourable circumstances, it is here a beautiful and very distinct tree, but unfortunately liable to suffer damage from late spring frosts. It is therefore only to be recommended for exceptionally mild and well-sheltered districts, and should only be planted in northern aspects, where its early growths will be shaded from the morning sun. Its habit of growth is broadly conical, the branches abundant, and thickly covered with broad flat leaves of from 2 to 2½ inches long, deep green above, and slightly glaucous below. It thrives best in rich loamy soil, with the ground well drained.

P. Webbiana (*Captain Webb's Indian Silver Fir*). —This noble tree is indigenous to the Bhotan and Himalayan mountains, where it forms vast forests at altitudes of from 9000 to 12,000 feet, and attains a height of from 50 to 80 feet. It was first introduced in 1822. It habit of growth is broadly conical, the branches horizontal, and regularly arranged round the stem. The leaves are from 1 to 2 inches long, thick and leathery, arranged on the shoots in two rows, of a deep shiny-green tint above, and glaucous beneath.

The cones which it occasionally produces in this

country have, while young, a fine violet colour, and are then very ornamental.

While this very beautiful tree, from its tendency to grow early in spring, is in most districts damaged by early frosts, and therefore cannot be recommended for general cultivation in this country, we would nevertheless recommend its being tried in such as are mild and thoroughly sheltered. Where it succeeds even moderately it is a noble object, and well worthy of the little trouble necessary for its protection from frost when making its growths in early spring. It should be planted in rich, deep, but dry soil.

PODOCARPUS (THE LONG-STALKED YEW).

In this group we have a large number of grand evergreen shrubs and trees, natives of Asia, Africa, and America, some of them very lofty, and many of them producing excellent and durable timber. Though botanically distinct, they are closely allied to, and resemble both in habits of growth and foliage, the various forms of our Yews, to which tribe some of the species are sometimes, though improperly, referred. Out of some forty or fifty species and varieties known to botanists, only two or three are sufficiently hardy for outdoor cultivation in Britain, and even these require to be planted in well-sheltered localities. All the sorts thrive best in shady aspects and in peaty soils, or loams rich in vegetable matter, and when well grown form distinct and interesting ornamental shrubs.

The following can be recommended as among the most desirable, and worthy of a trial where suitable conditions for their culture are available.

P. andina (*the Andes Podocarpus*), also known as *spicata*, is a native of high mountain-ranges in south Chile, where it is described as forming a thickly-branched, broadly conical shrub or small tree, varying in height, according to altitude or exposure, from 10 to 25 feet.

It is here a fine ornamental shrub, of rather slow growth, but very symmetrical in form, regularly furnished with branches from the ground, and well suited for planting as a single specimen on grass. The leaves are flat and broad, from $\frac{1}{2}$ to $1\frac{1}{2}$ inch long, and of a rich shiny-green colour.

P. koriana (*the Corean Podocarpus*), sometimes called *Taxus japonica*, is found wild in mountainous districts on the peninsula of Corea in China, and in similar situations in Japan, attaining heights of from 10 to 20 feet. It is also frequently met with in cultivation in both these countries as an ornamental shrub in town gardens, and in the vicinity of temples.

Though of remarkably slow growth here, it is

quite hardy, and forms a neat bushy shrub, with a habit of growth resembling that of the Irish Yew. The long upright branches are much divided into small branchlets, well furnished with linear-shaped leaves, from 1 to 2 inches long, of a bright glossy-green colour above, and faintly glaucous below. It is a superb plant for lawns or small gardens, and invaluable for geometric beds, or other arrangements where neat, compact, dwarf shrubs only are admissible. It succeeds best in deep, rich, loamy or peaty soil, and in damp rather than in dry situations.

P. nubigæna (*the Nubigean Podocarpus*).—This is a species of wide distribution at high elevations in the colder parts of Chile, on the Andes of Patagonia, in Valdivia, and on the island of Chiloe, and is described as a very beautiful tree, of from 20 to 30 feet in height. It was introduced in 1848. In this country it is a slow-growing plant, extremely fastidious as to soil and situation, and is only found to succeed in northern aspects well sheltered, and in cool rich loams. Where it does thrive, however, it is very ornamental, and worthy of a trial in collections of choice shrubs. The habit of growth is broadly conical, the branches closely covered with thick leathery leaves from 1 to $1\frac{1}{2}$ inch long, of a bright green colour above, and with a glaucous line on each side of the middle beneath.

PRUMNOPYTIS ELEGANS (THE GRACEFUL PRUMNOPYTIS).

This superb plant, so far as is known to us the only representative of the genus in cultivation, is a recent introduction from Valdivia, in South America, discovered by Mr Pearce, and described as a broad bushy shrub of only a few feet high, at altitudes of from 5000 to 6000 feet above the sea-level.

It is as yet only to be seen here in a young state, and presents the appearance of a miniature Yew. In habit of growth it is sharply conical, very profusely furnished with tiny branches from the ground upwards, the small Yew-like leaves, of a bright green colour, thickly covering the shoots.

The experience of the last two or three years proves that, though requiring a moderate amount of shelter, it is quite hardy in most localities, and at the same time, that it may be cultivated with the greatest facility in almost every variety of soil, with a preference, however, for light deep loam. It is doubtless a valuable acquisition, and will contrast well with the fine dwarf *Retinosporas*, and other compact-growing Conifers, in small garden beds, or in front rows of the taller forms, while no plant will be more effective for planting on and around rockeries.

RETINOSPORA (THE JAPAN CYPRESS).

The beautiful species of shrubs and trees which compose this genus are indigenous to Japan, where also, with their numerous varieties, they are extensively planted for ornamenting gardens and pleasure-grounds. In foliage, style of growth, and general appearance, they resemble the Cypresses, to which family, though separated by some peculiar botanical characters, they have evidently a close affinity.

Though little more than twenty years have passed away since the first representative of the group was introduced into European collections, there are already some eighteen or twenty species and varieties in cultivation, more or less distinct, but without exception eminently handsome; and even from the short experience we have had of them in this country, the great majority having been sent home since 1864, there is no doubt of their adaptability to our soils and climate, and that they are invaluable acquisitions to our list of hardy evergreen trees and shrubs.

In habit of growth they are for the most part dwarf and bushy, in some cases only rising a few feet from the ground, and are thus most useful for planting in miniature gardens or rockeries. Two or three of the species, however, attain the dimensions of trees, producing timber of such excellent quality, both as regards its durability and fineness of grain, that it is much used by the Japanese cabinet-makers for their most artistic work.

Like many of their congeners, the *Retinosporas* thrive best in a deep, rich, loamy soil, with the land sufficiently drained to prevent any accumulation of water at the roots; and though all the sorts have proved themselves hardy enough to withstand any amount of frost they have hitherto been subjected to in this country, they do best in a sheltered situation, or at least one in which they are not exposed to the full force of violent winds.

Each of the various species and varieties in cultivation has its own peculiar claims upon the attention of planters of ornamental evergreens, and all may be recommended, where space and circumstances admit, as desirable in even the most choice collections.

R. ericoides (*the Heath-like Retinospora*).—This is a pretty dwarf species, rarely found in its native habitats higher than from 4 to 6 feet.

It is much valued as a decorative plant by the Japanese, who cultivate it in their gardens, as well as in pots on balconies and terraces.

It is here a slow-growing, very compact bush, of a neat conical shape, clothed to the ground with tiny heath-like branches, densely covered with small leaves of a glaucous-green colour in summer, changing in winter to a bright violet purple. It is a fine rockery plant, and most useful either for winter bedding or for margins of shubberies, its peculiar colour contrasting strikingly with other plants of darker foliage.

It prefers a dry, airy, but sheltered situation, and a good rich soil; and, in order to allow it to develop its true character, should always be allowed sufficient space to be free from contact with other plants.

R. filicoides (*the Fern-like Retinospora*).—Like the preceding, this is a dwarf species, very hardy in this country, and forms an elegant bushy shrub, with abundance of small branches, divided into flat, frond-like branchlets. The leaves are small, but thickly set on the stems, and it has a rich, bright green colour, which it retains all over the year. It is altogether a most distinct and desirable plant, well suited for a rockery, or, indeed, any other situation where a miniature slow-growing evergreen is required.

Var. *nana aurea*.—This variety has a much dwarfer habit of growth than the parent, the branches are more slender, and freely variegated with a rich golden yellow. It is a neat little plant for the front of a small bed or a rock-garden.

R. filifera (*the Thread-branched Retinospora*).— This very beautiful and distinct species is described as rising to a height of about 50 feet in sheltered valleys at high elevations in Japan. It has long slender branches, gracefully drooping at the extremities, giving the plant a peculiarly soft feathery appearance. The leaves are short, very abundant, and of a bright green colour. In sheltered situations it has stood the test of the last few winters, and will doubtless prove a welcome addition to the list of hardy evergreens.

R. leptoclada (*the Flat-branchletted Retinospora*) is a dwarf form, never found in its native valleys higher than from 4 to 6 feet. It is a favourite pot-plant with the Japanese, and, as such, is met with very frequently in the gardens of Yeddo. In this country it is an exquisitely pretty shrub, of a close, sharply conical shape, the branches very dense, and divided into numerous short, flattened branchlets, each one resembling the fronds of a Fern or Lycopod. The foliage has a distinct silvery-grey colour, assuming a darker hue in winter. It is a superb wintergarden or rockery plant, and being quite hardy, and of easy culture, ought to be extensively grown, both for its neat form and its distinct colour.

R. lycopodioides (*the Club-moss Retinospora*), also a dwarf species, with a spreading rather than a conical habit of growth, the branches abundant, and divided into slender branchlets, densely clothed with small leaves, imbricated round the stem. The colour is deep green, equally so in winter as in summer.

It is an elegant and distinct-looking plant, well worthy of cultivation, and suitable for similar purposes as the other dwarf species, and, like them, though quite hardy, of slow growth.

R. obtusa (*the Blunt-leaved Retinospora*). — This magnificent species is found in various districts in Japan, particularly on the island of Niphon, where it is the principal forest-tree, and rises to heights of from 70 to 100 feet, with a straight arrow-like stem from 3 to 5 feet in diameter near the base. Its timber is fine-grained, capable of receiving a brilliant polish, and of a beautiful white colour—qualities so much appreciated by the Japanese that they regard it as sacred, call it the "Tree of the Sun," and use it in the construction of their temples and other religious buildings.

Though only introduced in 1850, it is already widely distributed; and enough has been seen of it to prove its thorough hardiness and free growth in almost every district of the country and in every variety of soil, and that it is an acquisition equal in importance to any of the grand importations of a similar kind from California.

In a young state its habit of growth is sharply conical; but as it advances, the branches become more spreading, and exhibit a tendency to assume the broad horizontal style characteristic of the tree in its native mountains. The foliage is of a rich dark-green colour, and the general appearance similar to that of some of the Arbor vitæs. It forms a fine lawn specimen, and is valuable as a contrast in mixed ornamental plantations.

Like all the rest of the species, it grows best when sheltered from the wind, and prefers a deep rich soil.

Of a considerable list of varieties, the following are most worthy of notice :—

Var. *aurea* has its green branchlets intermixed with others of a golden yellow.

Var. *argentea* differs only from the preceding in the colour of its variegation, which is silvery white. Both were found originally in gardens in Japan, as, indeed, were most of, if not all, the varieties in cultivation. These are much dwarfer than the species, but quite as hardy, and very effective in spring while making their young growths.

Var. *nana*.—This is a beautiful miniature form, of a compact, globular habit of growth, and very desirable for rockwork, or beds of the other dwarf Conifers.

R. pisifera (*the Pea-fruited Retinospora*), found wild in mountain forests on the island of Niphon, associated with *R. obtusa*, is a very handsome but much smaller tree, seldom rising higher than about 30 feet, producing timber of equal value to that species.

It is here thoroughly hardy, and grows more

rapidly than *R. obtusa*, forming a beautiful lawn specimen of a broadly conical shape, the branches very slender and shooting out from the stem in a horizontal direction. The foliage has a distinct, warm green tint on the upper surface, and is bright glaucous on the under. It thrives well under similar conditions with *R. obtusa*, to which it is a fine companion plant.

The following varieties can scarcely be too highly spoken of, their neat style of growth and brilliant-coloured variegations rendering them universal favourites. They are dwarfer than the species, but quite as hardy, and invaluable for flower-garden decoration:—

Var. *aurea*.—A beautiful little plant, of a compact, bushy habit, with its branchlets of a bright golden colour.

Var. *argentea* has most of the branchlets tipped with bright silvery white. Its habit of growth is similar to that of the preceding.

Var. *nana aurea*.—This is one of the dwarfest and neatest of Conifers, and prettily variegated, the branchlets appearing as if gilt with the purest gold.

R. plumosa (*the Plume-like Retinospora*).—This is a very distinct and interesting species, among the most recently introduced of the genus. The experience of the last four or five winters proves its perfect hardiness, and it is found to grow in ordinary soils quite as freely as the other sorts. In habit of growth it is very compact, with a blunt, round top. The branches are slender individually, but produced in great abundance. The foliage has a light-green or slightly glaucous tint. It is a superb dwarf shrub, not likely to rise higher than a few feet from the ground, and admirably suited for arrangements of neat, small-growing evergreens in geometric gardens.

Var. *aurea.*—A charming variety, with a fine golden tint intermixed with its green spray. Like its parent, it is of dwarf habit, and very effective, particularly in early spring and summer.

Var. *argentea.*—Another very beautiful form, differing only from the preceding in its terminal shoots being tinted with a silvery instead of a golden variegation.

R. **squarrosa** (*the Scaly-leaved Retinospora*).—This is a very distinct species, indigenous to the island of Kiusiu and the mountains of Sukejama in Japan, forming a dwarf shrub, rarely growing higher than about 8 feet.

It is here quite hardy if planted in a moderately sheltered situation, and is now widely distributed in our gardens, and highly valued for its compact, bushy habit of growth, and peculiar glaucous foliage. The branches are slender and very abundant, clothed with numerous small leaves. It is a remarkably pretty little plant, admirably adapted for associating with miniature shrubs in small garden-beds, to which, by its peculiar tint, it imparts a pleasing effect. It thrives best in a rich porous soil and in a sunny situation.

SALISBURIA ADIANTIFOLIA (THE MAIDEN-HAIR TREE.)

Of this beautiful genus this is the only species as yet known to botanists. It was first described and named by Linnæus as *Ginkgo biloba*, but subsequently changed by Smith and named in compliment to the distinguished English botanist R. A. Salisbury.

It is found not only wild but in cultivation as an ornamental and timber tree in many provinces of China and Japan. It is deciduous, and forms a graceful conical tree of from 50 to 100 feet high, with trunks of from 6 to 10 feet in diameter. The wood is soft and easily worked, but very fine-grained, and capable of receiving a high polish: it is extensively employed both in China and Japan for the finest cabinet-work.

It has been in cultivation here since 1754, and large, handsome examples of it are to be found all over the country. Its habit of growth is erect and bushy. The branches are somewhat irregularly disposed on the stem, horizontal, and much divided into small branchlets. The leaves, which are produced in great profusion, are about 1½ inch broad fan or wedge shaped, and have the appearance of the pinnules of a large Adiantum Fern ; they are, on both sides, of a pale-green colour.

Though of free growth in any rich garden-soil, and quite hardy as far as frost is concerned, it requires a sheltered situation, with the ground either naturally porous or well drained. Where it succeeds well, it gives a grand character to park scenery— the singular beauty of its curiously-formed leaves and its picturesque aspect commending it to all who have a taste either for elegance of form in foliage or for striking landscape effects. While young, it makes a pretty specimen for growing singly on a lawn.

Var. *macrophylla*.—This form differs from the parent only in having much larger and more divided leaves. It is a very desirable plant for a choice collection.

Var. *variegata*, a scarce but superb variety, with its leaves more or less striped with a bright golden variegation.

SAXE-GOTHÆA CONSPICUA (PRINCE ALBERT'S YEW).

This genus, which is as yet only represented by one species, was named in compliment to the late Prince Consort. It is indigenous to high mountain-ranges in Patagonia, from whence it was first sent to this country in 1846.

In its native habitats it forms a broad bushy shrub or small tree, varying in height from 10 to 30 feet, according to altitude and exposure. In general appearance it resembles some of the forms of the English Yew, to which it is nearly allied. It is, however, quite distinct, the leaves being longer in proportion to their breadth, irregularly arranged on the branches, and of a bright green colour.

Although found by its discoverer (Mr W. Lobb) at high elevations, very frequently occurring close upon the limits of perpetual snow, it cannot be classed among our hardiest Conifers. It is rarely, however, killed outright by our severest frosts, and is successfully cultivated in mild, well-sheltered localities when planted in rich, deep, well-drained soil, and in aspects shaded from the direct influence of the morning sun in winter and early spring, which after hard frost is almost always fatal to the young shoots. In a healthy vigorous state it is a very handsome and interesting plant, and well deserves a trial in any suitable situation.

SCIADOPYTIS VERTICILLATA (THE UMBRELLA PINE).

This curious and very handsome Conifer, the only species of the genus at present known, or at least in cultivation in British gardens, is indigenous to mountains in Japan, and is there also extensively planted in pleasure-grounds and in the vicinity of temples. It is described as the "finest Conifer in Japan," forming a compact, conical tree, with a straight stem and horizontal branches, and attaining in its native habitats generally heights of from 60 to 70 feet, very old specimens being occasionally found considerably over 100 feet.

The leaves are from 2 to 4 inches long, flat, and tapering to an obtuse point, of a light-green colour, and arranged in whorls of from 30 to 40 at the ends of the shoots, each group presenting the appearance of an inverted umbrella.

Since its introduction in 1855, it has been widely distributed over the country; and though proved to be quite hardy, and to succeed well in many cases, it is on the whole a very partial plant, slow-growing in the extreme, and not likely ever to attain dimensions larger than those of a moderate-sized shrub. It is nevertheless so very handsome, and so distinct from any of its congeners, that it is indispensable in any collection of small evergreens. It seems to thrive best in a deep, rich, loamy soil, and should always be planted in a well-sheltered situation.

SEQUOIA SEMPERVIRENS (THE CALIFORNIAN RED-WOOD TREE).

This is the only species of the genus, although several writers maintain that the *Wellingtonia* should be associated with it. It was originally referred to *Taxodium*, probably from the great similarity of its foliage to that of the deciduous Cypress, but was afterwards very properly separated into a distinct genus by Endlicher, whose arrangement is now adopted by the best authorities.

It was first discovered by Menzies on the northwest coast of America in 1796, and by subsequent explorers in great abundance all over Upper California, from whence it was introduced to Britain in 1825.

It is an evergreen tree of gigantic proportions, attaining heights only surpassed by its near ally the *Wellingtonia*, individuals being frequently met with from 200 to 270 feet high, with circumferences of from 20 to 40 feet, and straight trunks clear of branches to heights of from 60 to 70 feet. It is called by the settlers "the giant of the forest." Douglas remarks of this tree : " But the great beauty of the Californian vegetation is a species of *Taxodium*, which gives the mountains a most peculiar, I was almost going to say awful, appearance—something which plainly tells us we are not in Europe."

Where this tree succeeds in this country it is a magnificent park specimen, with a broad, conical habit of growth, and with leaves and branches very much resembling the Yew; and it is found to grow with remarkable rapidity and luxuriance in almost all soils. Unfortunately, however, its tendency to grow late in autumn, through which it so frequently loses its leading shoot and gets otherwise damaged by frost, must always be a great barrier to that popularity which its other merits as a decorative tree would insure it among landscape-gardeners. It is nevertheless handsome enough to be worthy of a trial in even a limited collection; and in most cases where the circumstances are even moderately favourable, the result will be satisfactory. A well-drained soil, not too rich, an aspect somewhat shady, and a situation well sheltered from violent winds, should always be chosen for this tree, as being most likely to insure its successful cultivation.

Var. *Lawsoniana*.—This variety is of recent introduction to our collections; it differs from the species in having a more compact habit of growth, with short, stiff branches, and much smaller leaves. It is also of slower growth, and forms a neat specimen shrub, valuable for lawns or arrangements of the neat-growing miniature Conifers.

TAXUS (THE YEW).

Though limited to only two or three species, and even these regarded by some botanists as specifically identical, this genus is remarkably rich in distinct and useful varieties—many of them possessing in such a high degree those qualities so desirable in outdoor decorative trees or shrubs, that they form a prominent feature in almost every ornamental plantation. It has a wide geographical distribution, the various forms being found more or less abundantly over North America, the temperate parts of Asia, and most of the countries of Europe, generally at high elevations, frequently extending to 4000 feet above the sea-level, but thriving best in sheltered valleys, where they attain in some cases the dimensions of large timber-trees.

The wood, which from its strength and elasticity was extensively employed by our ancestors before the invention of gunpowder in making bows, is now highly valued for artistic cabinet-work, turning, and carving. It is of a beautiful brownish-red colour, sometimes nearly white, frequently veined, very hard, close-grained, and susceptible of receiving a high polish.

All the sorts are of slow growth, particularly after the first few years, or after they attain heights of from 15 to 20 feet—their tendency then being to increase in breadth rather than in height.

Few trees are less fastidious in regard to soils and situations; and handsome specimens are to be found in every district of the country, growing and thriving in almost every variety of soil. There is no doubt, however, that they prefer a deep rich loam, with a subsoil cool and moist, and a situation moderately sheltered and shady rather than exposed to the full rays of the sun. They are, indeed, among the few evergreens that succeed well in the shade of high trees, not only growing well in such circumstances, but developing the peculiar dark glossy green of their foliage to the greatest perfection.

T. baccata (*the English Yew*), the European form of the genus, is indigenous to Britain, and also found abundantly on most of the great mountain-ranges of the Continent, including the Alps, the Apennines, the Pyrenees, and the Caucasus, at elevations of from 1000 to 4000 feet.

This shrub or small tree, even when at full maturity, rarely exceeds 30 or 40 feet in height, with a trunk remarkably thick in proportion to its length, in some instances exceeding 50 feet in circumference at the base, the long spreading branches so abundant, and so densely clothed with branchlets and foliage in isolated specimens, that the stem is completely covered from the ground upwards.

Though cultivated in British gardens and pleasure-grounds, in churchyards and cemeteries, from time immemorial, this grand evergreen is still as popular as ever—indispensable among ornamental plants, occupying a place peculiarly its own, and forming the most effective contrast with most other shrubs, whether in groups or as single specimens; while the deep sombre green of its foliage and its stiff formal aspect suggest to every mind its singular appropriateness as an adornment to the resting-place of the dead. A popular writer, himself now reposing under "the Yew-tree's shade," aptly says, "For the decoration of places of burial it is well adapted, from the deep and perpetual verdure of its foliage, which, conjointly with its great longevity, may be considered as emblematical of immortality."

"The Yew," remarks another author, "is celebrated both for its military and superstitious uses in England. These trees were anciently planted in our churchyards either to supply the parishioners with bows or to protect the church from storms. In every nation it is considered the emblem of mourning. Its branches were carried in funeral processions by the friends of the deceased. The Yew has thus partly acquired an almost sacred character."

Having naturally a dense twiggy habit of growth, and as it may be clipped or cut into almost any shape with the greatest impunity, the English Yew has few equals as an evergreen garden-hedge plant, and as such it has been long extensively employed; and in the day, now happily passed away, when it was fashionable to adorn gardens with shrubs cut into architectural forms as well as into those of animals, and even man, its patience under the knife was amply taken advantage of, producing some of the most grotesque and intricate designs, with a solidity and sharpness of outline superior to either Juniper or Boxwood.

From a long list of varieties we recommend the following as being the most distinct and ornamental:—

Var. *fastigiata*.—This is the well-known Irish Yew now so frequently planted in cemeteries, a shrub so different in foliation and general appearance that it is difficult at first sight to identify it with the species. The branches, which are very abundant and divided into numerous short branchlets, instead of spreading out horizontally, are fastigiate or upright, like those of the Lombardy Poplar, but are so compact that the plant looks like a solid column. The leaves are of the darkest sombre green, and disposed either in tufts, or scattered irregularly round the branches; while the berries, instead of being round, are distinctly oblong. Seeds saved from this curious form rarely if ever yield anything else than the species, proving that it is neither more nor less than one of these strange sports which occur from time to time among plants found wild in various regions of the world, and growing under widely different conditions as regards climate and soils.

It was first found in the year 1780 on a mountain near Benoughlin (Lord Enniskillen's estate) by a tenant, who brought it to the gardens at Florence Court; and it is believed that from this plant, or plants, all those now in cultivation originated. Two beautiful variegations of this grand variety have been introduced within the last few years to our gardens—the one with its green leaves intermixed with those of a white variegation, the other with bright yellow, and named *argentea* and *aurea* respectively. Both, particularly the latter, are great acquisitions; and though as yet comparatively scarce, and rarely met with in collections, will doubtless be in greater demand as soon as their merits become better known. Like their parent, they are remarkably hardy, and valuable for lawns, terraces, margins of walks, or as contrasts in mixed arrangements of ornamental shrubs.

Var. *Cheshuntensis.*—This variety resembles the Irish Yew in its close, upright habit, but is more conical, with the branches more diffuse. It is said to have originated from seed saved from that sort. It is a very desirable lawn plant.

Var. *Dovastonia.*—This is one of the finest of the sports from the species; the leaves are much larger, of a dark glossy green, and the branches long and pendulous. When the main stem of this plant is trained carefully up, and the side branches allowed to droop, it is a most attractive lawn specimen. It is sometimes grafted on stems of the common sort, and forms a neat standard weeping shrub, with branches shooting out horizontally and drooping at the points.

Var. *glauca*, sometimes called *Blue John*, is a very striking variety, of free, vigorous growth, and of a more spreading habit than the species, but differing more particularly in the colour of the leaves, which are deep green on the upper surface, but on the under of a peculiar bluish grey, while the bark on the young shoots is of a rusty brown: it is a very interesting and desirable plant.

Var. *fructo-lutea* differs from the species in no other respect than that its berries, instead of being red, are of a bright golden-yellow colour: it is very effective when in fruit.

Var. *aurea*.—The Golden Yew is one of the most beautiful of variegated shrubs, having a brilliancy of colour which renders it invaluable as a contrast in mixed groups: this variety is sometimes grafted on the top of both the common and Irish Yew, in each case with grand results.

Var. *elegantissima*.—This is another variegation, rather straw-coloured than golden, very distinct from the preceding, and equally valuable, whether on its own roots or grafted on the English or Irish sorts.

Var. *ericoides.*—This variety differs only from the species in its being much more slender in all its parts. It is an interesting tiny shrub, and suitable for front rows or small beds.

Var. *erecta.*—This is a neat, close-growing, conical-shaped shrub, densely clothed with leaves, much smaller than the species, and forming an attractive lawn plant.

Var. *erecta aurea.*—This is a very beautiful variegation, having blotches of gold intermixed with the green leaves.

Var. *nana,* or *Foxii,* is a very dwarf form, rarely rising above 2 feet from the ground: the branches are slender and spreading, the leaves are very small but numerous. It is a fine shrub for planting on and around rockeries, or for the margins of small beds.

T. adpressa (*the Flattened or Creeping Yew*), by some authors believed to be only a variety of *baccata,* is a native of mountains in Japan, from whence it was sent home to this country a few years ago. In its native habitats it is rarely found higher than about 7 or 8 feet, forming a spreading, flat-headed shrub, with numerous branches, densely furnished with short branchlets, thickly clothed with short, flat, dark-green leaves. On its own roots in this country it presents very much the appearance thus described, and is only really useful for rockeries or small beds, in which situations it is very distinct and interesting. Grafted, however, as it now very often is, on stems of the English or Irish sorts, it is one of the prettiest of weeping shrubs, and ought to be extensively planted in choice collections. Of this species, the following variety deserves attention :—

Var. *stricta*, this is a greater favourite with planters than the species itself, having a more upright and vigorous habit of growth: when trained to a single stem it forms a neat dwarf front row or lawn shrub.

T. canadensis (*the Canadian Yew*), like the preceding species, is regarded by some writers as only a variety of *baccata*. It is found wild in considerable abundance in Canada and the United States, growing in shady situations, from whence it was first introduced about 1800. It grows about 4 feet high, and has a spreading bushy habit, with short, pale glossy-green leaves abundantly clothing the branches. The bark is of a dark-brown colour; the berries, which are smaller than those of *baccata*, are of a bright coral red, rendering the plant very attractive when they are ripe, as they are generally produced in great abundance. It is a distinct and interesting dwarf shrub, quite hardy, and of free growth when planted in a moderately moist, sheltered situation.

Var. *Washingtonii*, has its leaves prominently variegated with a bright golden yellow.

T. Lindleyana (*the Californian Yew*), by some botanists called *Baccata americana*, found in valleys and on river-sides in Northern California, is described as a handsome tree of from 40 to 50 feet high, and from 4 to 5 feet in circumference near the ground. The branches are long, slender, somewhat pendulous, and covered with a yellow or light-brown bark; the leaves, which are produced in great abundance, are very similar in size and shape to those of *baccata*, but of a light-green colour. In a young state this plant is very distinct and pretty. It seems to be quite equal to our climate, and promises to be a useful and interesting acquisition to our collection of ornamental shrubs.

TAXODIUM DISTICHUM (THE DECIDUOUS CYPRESS).

This, the only species of the genus hardy enough to warrant its being recommended for outdoor cultivation in Britain, is an eminently beautiful deciduous tree, introduced from North America so early as 1640. It is found wild over a vast surface of country, extending from Delaware to Florida, and close to the sea in Virginia and Maryland, chiefly in low swampy ground, and on the moist banks of rivers, forming immense forests; those on the banks of the Mississippi stretching almost without interruption for hundreds of miles, and often attaining heights of from 80 to 150 feet. A curious circumstance connected with this tree is that in its native habitats, when it reaches a height of about 30 feet, the roots begin to throw up large hollow protuberances above the soil: these are known as "Cypress knees," and are used by the settlers for bee-hives. They are always largest in low marshy ground.

Though rarely seen in this country higher than about 30 feet, there are few more graceful trees in cultivation; its warm green tint, and light feathery appearance in summer, along with the brilliant red of its decaying leaves in autumn, presenting a striking and pleasing contrast with almost every other tree with which it can be associated, and producing quite a pictorial effect. While young it has a close, bushy, conical habit of growth, branched to the ground, the lower branches afterwards dying off, and the tree gradually assuming a flat-headed shape. In either state it is a beautiful object on the lawn, or in kept grounds, and ought to be far more freely planted than it has hitherto been. It grows with great luxuriance in low-lying damp situations, but in such, is extremely liable to suffer injury from frost; it should therefore be planted in open, airy, but sheltered places, where it will be protected from high winds, and induced to ripen its wood thoroughly before winter sets in, as well as prevented from starting into growth too early in spring, the frosts of which often nip its young buds.

Var. *pendulum*, called by some botanists *Taxodium sinense*, and by others elevated into the type of a new genus under the name of *Glyptostrobus pendulus*—the majority, however, retaining it under this arrangement—is a distinct and handsome form, introduced in 1845 from China, where it forms a small tree of about 20 feet. It differs from the species chiefly in having smaller leaves, and in its drooping branchlets. It is an exquisitely beautiful plant; and though scarcely so hardy as the American form, is worthy of a trial in the more favoured localities.

THUJA (THE AMERICAN ARBOR VITÆ).

This is a small but very interesting genus of evergreen trees, natives of North America, some of them attaining great dimensions, and highly valued for their timber. They are all quite hardy here, very ornamental, and of remarkably free growth in almost every variety of soil.

The different species and varieties possess, in a greater or less degree, that upright, densely branched, conical habit of growth so characteristic of the Cypresses and some of the Junipers, but so distinct in appearance, and so uniformly handsome, that they are rarely absent even in small collections of decorative shrubs.

T. gigantea (*the Tall Arbor vitæ*).—This species, hitherto known and distributed under the names *Menziesii* and *Lobbii*, is now ascertained to be the true *gigantea* first described by Nuttall in his 'Plants of the Rocky Mountains;' the plant erroneously so named being the *Libocedrus decurrens*—a tree that, even under the most favourable circumstances, in its native woods rarely rises higher than 50 feet, whereas *gigantea* grows to a height of nearly 200 feet.

This grand tree is found wild at altitudes of from 4000 to 5000 feet, over immense tracts on the north-west coast of America, and in California, from whence seeds were first sent home by Jeffrey in 1854.

Of this species in its native habitats Nuttall remarks: "This is one of the most majestic trees west of the Rocky Mountains, attaining the height of 60 to 170 feet, or even 200 feet, and being 20 to 40 feet in the circumference of the trunk. On the shores of the Pacific, where this species is frequent, it nowhere attains the enormous dimensions attributed to it in the fertile valleys of the Rocky Mountains, towards the sources of the Oregon. We seldom saw it along the coast more than 70 to 100 feet in height."

Though as yet only seen in this country as a large

shrub, or at most but beginning to assume the tree form, and chiefly confined to parks and pleasure-grounds, it has been widely distributed and extensively planted, so that its thorough hardiness, freeness and rapidity of growth in almost every district and variety of soil, have been amply demonstrated; and there is every reason to believe that it will yet come to be regarded as one of the most valuable of British forest-trees.

In a young state, and as seen in our pinetums and parks, it is one of the most beautiful of its tribe, having a handsome conical habit of growth, clothed to the ground with long graceful branches, much divided into feathery branchlets, of a shining, warm green colour, which is maintained all over the year.

Of varieties, many of which may be detected in almost every lot of seedlings, the most distinct and constant is—

Var. *Craigiana.*—This fine form was raised from the seeds sent home by Jeffrey, and differs from the species chiefly in having a more open habit of growth, with the branches slightly pendent, and turned up at the extremities. It is quite as hardy, and of as free growth, as the parent, and is a grand specimen either for the lawn or park.

T. occidentalis (*the Western Arbor vitæ*) is indigenous to, and occurs in great abundance over, a large area in Canada and the United States, where it is commonly called the "White Cedar," and from whence it was first introduced into Britain about 1596. It is usually found growing in low sheltered swamps, and on the moist banks of rivers, where the soil is of a peaty or rich alluvial character, rising to heights of from 30 to 50 feet.

The timber being close-grained and remarkably durable, as well as light and easily wrought, is extensively used in America for fencing, house-building, and a variety of other purposes.

As an ornamental shrub, for which it is alone cultivated in this country, it has long been highly popular; its symmetry of outline and profusion of graceful plumy branches, along with its beautiful light-green summer's tint, render its presence ever welcome, either in mixed groups of shrubs, or as single specimens on the lawn; and though it assumes a somewhat sombre russet brown on the approach of winter, it is even at that season, as a contrast to the lively greens of many of the Cypresses, strikingly pleasing and effective.

From its dense bushy habit and facility of growth in almost all soils and situations if moderately moist, this species is well adapted for forming garden screens and ornamental hedges, which, if the operation is performed in early summer, may be freely pruned or trimmed into any shape.

The following varieties are distinct, very handsome, and as hardy as the species:—

Var. *ericoides*, or *Elwangeriana*, is a neat, dwarf,

bushy plant, with a great profusion of tiny heath-like branches; is useful for planting in front of groups of the larger shrubs, for winter-bedding, terrace-vases, and as single specimens on lawns of small extent.

Var. *compacta* differs from the species in its branches being more compressed and produced in greater abundance: it is one of the prettiest of cone-shaped lawn specimen plants, and very useful for planting on terraces or edges of walks.

Var. *variegata aurea*, has some of the branchlets tinted with a light-yellow variegation, and though of slower growth and more slender than the species, is a very desirable shrub.

Var. *vervæniana.*—This very interesting variety, which originated a few years ago on the Continent, has a dwarfer habit of growth and much more slender branches than the parent: it is valuable for spring bedding on account of the brilliant yellowish hue of its branchlets.

Var. *pendula.*—This is a curious form, differing from the species in its branches being drooping, with the branchlets clustered at the extremities: it is interesting as a rockery plant.

Var. *lutea*, or *George Peabody*, is a new and as yet a scarce plant. The branchlets are prominently tipped with the brightest golden variegation.

T. plicata (*the Plaited-leaved Arbor vitæ*) is found

wild on the western shores of North America, particularly at Nootka Sound, growing in deep alluvial soils to heights of from 20 to 30 feet. This fine species was introduced into Britain about 1769, and has proved itself to be a thoroughly hardy and free-growing ornamental shrub, of a compact, conical habit of growth, abundantly furnished with short, stout, horizontal branches, much divided into thick branchlets disposed in regular rows ; and these overlapping each other, give the plant that peculiar appearance which doubtless suggested the name *plicata*, or plaited. In summer the branches are of a light, slightly glaucous green, changing in winter to a rich brownish tint.

Though one of the commonest of our hardy coniferous shrubs it is well worthy of a place of honour among the rarest and most select, and handsome enough for a lawn, or any other site where a distinct symmetrical evergreen of moderate size is required.

It thrives best in a deep, cool, moderately moist soil, and should always be allowed plenty of space to develop its branches on every side. The following are distinct and useful varieties:—

Var. *minima* is a very dwarf form, of a dense, bushy habit, forming a neat little shrub for planting on and about rockeries, in small beds, or in terrace-vases; and being quite as hardy as the parent, thrives in very exposed situations.

Var. *wareana*.—This form, by some writers classed as a distinct species, differs chiefly from the type in having a more robust habit of growth—a distinction, however, that is not always very obvious, and which seems to depend very much on the soil or other circumstances under which the plants are grown. The same remark applies also to *siberica*, which so closely resembles this species as to justify the belief that it is a mere variety.

THUJOPSIS (THE BROAD-LEAVED ARBOR VITÆ).

This new and as far as is yet known small genus, is so named from the resemblance of the species of which it is composed to their near allies, the American Arbor vitæs. They are all natives of Japan, and though only introduced to our gardens at intervals during the last twenty years, enough has been seen of them to prove their adaptability to our soils and climate, and their great value as ornamental shrubs and trees.

T. dolobrata (*the Hatchet-leaved Thujopsis*).—Indigenous to high but sheltered valleys in Japan, and frequently met with in cultivation in that country as well as in China, both as an ornamental tree and for its timber, which, from its closeness of grain and durability, is much prized, and used extensively for a great variety of purposes. It is described as a handsome, broadly conical tree of some 80 feet in height, with spreading branches, drooping at the points; the branchlets numerous, much compressed, and clothed with flat scale-like leaves, regularly imbricated, of a bright glossy green on the upper, and silvery on the under surface.

This grand species was first sent to Britain in 1854, and having since been widely planted, and exposed to the rigours of our winters in the open air without sustaining damage, its perfect hardiness is now undoubted; and though of very slow growth while in a young state, it seems to improve in that respect with age; and there can be no question but that it will come to be regarded as one of our indispensable ornamental trees. As seen here it is a bushy shrub, broad in proportion to its height, the main stem very little in advance of the side branches for the first few years, but afterwards gradually taking the lead, and the plant assuming a conical shape. It makes most progress in strong loamy or peaty soils, moderately moist, and seems to prefer a shady aspect.

Var. *variegata*, sent home from Yeddo, in Japan, by Mr Fortune in 1861, is a pretty variety, with the branchlets more or less tinted with pale yellow. It is a very popular plant with the Japanese for the adornment of their residences and temples. It is here quite as hardy as the species, and one of the finest of our ornamental evergreens.

T. lætevirens (*the Lycopod-like Thujopsis*), sent home a few years ago by Mr Veitch from Japan, is a dwarf shrub, rarely found even in the most favourable circumstances to exceed 10 feet in height.

This beautiful little plant has a dense, broadly conical habit of growth; and though said to be specifically different, is in general aspect suggestive of a miniature form of *dolobrata*. It is, however, so distinct, that the one can never be mistaken for the other. The branches are very slender, and are divided into numerous flat fan-like branchlets, abundantly clothed with tiny neatly-cut leaves of a warm green colour, arranged with the utmost regularity, giving it a remarkable resemblance to a tree Lycopod.

Like *dolobrata*, it is of slow growth, and though quite hardy, should always be planted in a sheltered situation. It prefers a rich, deep, and moderately moist rather than a dry soil. As a neat bushy shrub for rockeries, or small beds where only plants of such habits are admissible, it has few superiors; and as its merits become better known, it will doubtless be extensively planted in such places.

T. Standishii (*Mr Standish's Thujopsis*), named in compliment to Mr Standish of the Ascot Nurseries, was sent home from Japan in 1861 by Mr Fortune, who discovered it near Yeddo.

This very handsome species is aptly described as

having an appearance "midway between a Thujopsis and an Arbor vitæ." It has an erect conical style of growth, densely furnished with slender branches, much divided into flat Lycopod-like branchlets, drooping at the points. The leaves, which in shape and arrangement on the stems are very like those of *dolobrata*, but much smaller, are light or yellowish green above, assuming a deeper tint in winter, and slightly glaucous below.

It is here thoroughly hardy, and seems to be quite at home under similar conditions to the other species. Though only seen in this country in a small state, and as yet comparatively little known, there is much in its appearance to warrant the belief that it will make a grand specimen ornamental shrub, and to recommend its being planted where such is desirable.

TORREYA (THE FŒTID YEW).

This small genus, named in honour of the late Professor John Torrey of New York, one of the authors of the 'North America Flora,' is closely allied to the Yews, and is composed of small evergreen trees, natives of North America, Japan, and the north of China. They are remarkable for the peculiarly disagreeable rue-like odour they emit when bruised or burned,—hence their popular name.

Though very ornamental, with one or two exceptions, they are unfortunately scarcely hardy enough to be very generally planted as shrubbery plants in Britain; and though specimens of all the species are to be found in exceptionally mild localities, they suffer injury to a greater or less extent in most winters, and present an appearance, even at their best, the reverse of handsome.

The two following are noted as the hardiest of the genus, and as likely to succeed when well sheltered; they are at least interesting enough to be well worthy of a fair trial where a favourable situation is available.

T. nucifera (*the Nut-bearing Torreya*). — This species is found wild on mountains in the north of Japan, particularly on the islands of Niphon and Sikok, where it occurs in great abundance as a large shrub or tree of from 20 to 30 feet in height, and is also cultivated all over Japan, its nuts producing an oil useful for culinary purposes.

It is here a compact, broadly conical bush, with numerous horizontal branches divided into short branchlets, thickly clothed with Yew-like leaves of a dark glossy-green colour above and slightly glaucous underneath. Where this plant succeeds, it forms an extremely pretty and interesting specimen, very desirable as a variety among other dwarf or slow-growing Conifers.

T. myristica (*the Californian Nutmeg*).—This is a fine species, indigenous to the Sierra Nevada

Mountains in California: introduced in 1848. It is said to be a small bushy-headed tree with spreading horizontal branches, growing to heights of from 20 to 40 feet. The timber is of a light-yellow colour, heavy, fine-grained, and is said to resist the attacks of insects better than that of any of the other pines.

Since its introduction to Britain it has been widely distributed in almost every district; and while it undoubtedly requires to be well sheltered, it has proved itself to be the hardiest of the genus, and fine specimens are frequently to be met with growing as freely as the common Yew. Its habit of growth is sharply conical; the branches, which clothe the stem from the ground upwards, are divided into short stiff branchlets; the leaves are shaped like those of the Yew, and from 2 to 2½ inches long, of a yellowish-green colour above, and of a paler tint underneath. It is an elegant lawn plant, and valuable as a contrast with others of darker tints and denser habits of growth.

WELLINGTONIA GIGANTEA (THE MAMMOTH TREE).

This well-known and highly ornamental evergreen tree, the only species of the genus, was named by Dr Lindley in compliment to the late Duke of Wellington; the propriety of this name has, however been questioned, some authorities contending that it is not generically distinct from *Sequoia*, and that it can only be correctly referred to that genus; while some American botanists, assuming it to be a new type, say that, as it is the largest tree in the world, and a native of America, it should be called, after their great man, *Washingtonia;* others again would have it named, after the continent itself, *Americanus*. Though inclined to believe that its true place is among the Sequoias, we have retained it under its original name, as that by which it is best known all over the world.

It is a native of the slopes of the Sierra Nevada, in Upper California, occurring in great abundance, sometimes in groves, and at other times scattered through the forest, over a vast tract of country, at altitudes of from 4000 to 7000 feet. Though discovered and described by Mr Douglas in 1831, it was only introduced in 1853, seeds having been sent home in the summer of that year by Mr William Lobb, the successful botanical explorer.

The marvellous reports, first from Douglas, and amply confirmed by subsequent travellers, of the colossal size attained by this tree in its native habitat, compared with which the most gigantic hitherto known are mere dwarfs, combined with the strong probability of its being hardy enough for our climate, rendered its introduction an event of more than ordinary importance, not only to practical arboriculturists, but to all who were in any way interested in ornamental trees; and as might be expected, it was rapidly distributed and extensively planted throughout the country. The "Mammoth Grove," as the Americans call the locality where the *Wellingtonia* was first met with, and where the largest specimens are found, is situated about 200 miles from San Francisco, near the head-waters of the Stanislaus and San Antonio rivers, in a sheltered valley some 4600 feet above the level of the sea. In this spot, within an area of about 50 acres, there are still standing from eighty to ninety specimens, the average height of which is 300 feet, with a circumference of 80 feet. These dimensions are very much exceeded in several individuals, one of the largest being 365 feet high, and 93 feet in circumference; while the "Father of the Forest," blown down ages ago, has a circumference of 110 feet at the ground. In falling, this giant had come in contact with some of his neighbours, and his trunk was broken off at a height of 300 feet; at this point he is 40 feet round, so that although the top has been long since destroyed, reckoning from the average taper of the others, it is not difficult to believe that he must have stood fully 450 feet high.

In an account of a visit to the Mammoth Grove, which appeared in 'Blackwood's Magazine' a few years ago, the writer remarks: "We wandered about the Grove for several hours, amid a scene of wonders, the mere description of which we should have laughed at as a traveller's tale. There are about one hundred trees of this species, of every age and size, intermingled with various kinds of Pine, Yews, and deciduous shrubs, and all standing within an area of about fifty acres.

"The younger ones are singularly graceful and handsome; but those of mature growth—a few thousand years old, perhaps—are a little withered at the top. The enormous trunks are bare and branchless for from 100 to 130 feet, and the boughs seem small in proportion to the central stem.

"The effect of the mighty columns rising thickly round, and towering on high, some burnt hollow, in whose cavities a company of soldiers might almost find shelter—others uninjured, solid and massive, the largest and oldest of living organisms on earth, monuments of ages past, when there were giants in the land—is almost awsome; and as we walked about, pigmy and insignificant, we half expected to see the strange forms of extinct giants of the animal world, the mammoth or the mastodon of ages still more remote, come crashing through the timber, or the pterodactyl winging its way amongst the colossal vegetation. There stood the 'Mother of the Forest' withered and bare, her full height 327 feet, her girth 78 feet without the bark, for this had been removed from 116 feet of the lower portion of the trunk, and the scaffolding erected for the purpose still stood round the tree. This outer shell thus removed is now put up in the Crystal Palace at Sydenham."

Various estimates, founded upon the annual rings of one of the largest, which was cut down a few years ago, have been formed as to the ages of these mammoths, and it is now generally believed that they must have seen the summers and winters of at least 3000 years.

The bark is more or less furrowed, giving the trunk the appearance of being made up of fluted columns, and in the old trees varies in thickness from a few inches to nearly 2 feet. The wood very much resembles Cedar, being light, soft, easily wrought, and of a pale-red colour.

In our pleasure-grounds—to which the *Wellingtonia* is as yet almost entirely confined—it is one of the most popular of ornamental trees, unsurpassed for planting in rows, either by themselves or associated with other varieties, in long wide avenues; forming magnificent specimens singly on lawns, and indeed worthy of being selected for any position where an effective symmetrical tree is desirable.

Though perfectly hardy, and capable of resisting any amount of frost it is likely to be subjected to in this country, it requires to be planted in a situation sheltered from violent cutting winds, of which in common with many others of the finest of the Pine tribe, it is most impatient, and from which it frequently suffers severely.

It is found to grow with the greatest luxuriance in rich deep soils, with a subsoil cool and moderately damp without being wet. Where the soil is composed of porous sand or gravel, and liable to extreme drought in summer, it makes little or no progress, and never becomes a fine specimen.

In favourable circumstances it assumes a neat conical form, with a straight robust stem densely clothed with branches from the ground upwards. The leaves, with which the branchlets are densely covered, are short, needle-shaped, and of a light-green colour.

Of several varieties in cultivation, the following is the most desirable :—

Var. *aurea variegata.*—This fine form originated in the Lough Nurseries, Cork, a few years ago, and is now well known as one of the handsomest of our variegated Conifers. The green branchlets are intermixed with others of a rich golden yellow; and as it has the neat conical habit and free growth of the parent, it will doubtless be much appreciated, and largely planted in lawns and shrubberies among the choicest of its tribe.

WIDDRINGTONIA CUPRESSOIDES (THE AFRICAN CYPRESS).

This is a small, dense-growing, evergreen shrub, of recent introduction from mountains in Northern Africa. There are several species known and described by botanists, but with this exception they are too tender for our climate. It is said to grow to a height of about 6 feet in its native habitats. As seen here, it is a remarkably pretty plant, with a general appearance suggestive of some of the *Retinosporas*, to which, in all probability, it is nearly allied. It is of easy culture in most soils, and has stood the test of the past few winters without damage from frost. Being of slow growth, and very compact in its habit, it will be a useful plant for associating with the other dwarf Conifers in small beds or edgings, where its fine glaucous hue will afford a pleasing contrast.

RHODODENDRONS

AND OTHER

AMERICAN OR PEAT-SOIL SHRUBS.

———◆———

Although the term American Shrubs has long ceased to be a correct one in its indiscriminate application to the great family of hardy peat-soil plants cultivated in this country, seeing that America contributes a comparatively small proportion of its numerous members, it is nevertheless still used conventionally to designate that group which includes the Rhododendron, Azalea, Andromeda, Erica, and the many other genera which can only be cultivated in peat, or in such soils as contain a large amount of the elements of which it is composed; and this, perhaps, not altogether inappropriately, as commemorative of the fact, that from that continent were first introduced some of these grand representatives, which, notwithstanding the rivalry of subsequent introductions from nearly every other quarter of the globe, still maintain their position among ornamental shrubs, and are as popular, and as extensively planted as ever.

The natural order *Ericaceæ*, to which, with few exceptions, these plants either belong, or with which they are closely allied, is remarkable for its wide geographical distribution, stretching to the utmost limits of ligneous vegetation in both hemispheres; and is peculiarly interesting to horticulturists, from the great diversity of forms and habits of growth found among the various genera and species of which it is composed; while the uniform beauty of flower, and in nearly every case elegance of foliage, give it a value in decorative planting unsurpassed by any other known family of plants.

Before proceeding to notice more particularly some of the more prominent and desirable of the genera and species, we may premise that, so far as culture and management is concerned, the suggestions we intend to offer in reference to Rhododendrons, which because of their great importance will be placed first in our list, apply generally to all the genera and species in the group—their requirements, with some special modifications, which we shall advert to in dealing with the various sorts, being very similar.

RHODODENDRONS (THE ROSE BAY).

"Rhododendrons, the pride of European gardens, as they are of their native wilds"—so wrote the great Loudon nearly fifty years ago, when very few species or varieties were either known or cultivated in Europe; and even many of these, though deeply interesting and worthy of careful cultivation, by no means conspicuous or striking as decorative plants. The gorgeous Indian *R. arboreum* was but recently introduced, and had not yet bloomed in this country, though wondrous things were said of its tree-habit and dark-crimson blossoms; and as yet the still popular *R. Ponticum, Catawbiense, Caucasicum, maximum, ferrugineum, and hirsutum,* and a few of their varieties, very nearly made up the list of what we may term showy sorts, though all were worthy of the high praise which was accorded to them.

Since then, it is scarcely necessary to say, a vast, and in some respects a remarkable, improvement has been effected, both in point of variety and quality, through the introduction of new species from their native habitat, as well as by the labours of hybridisers; so that, instead of the two or three sorts which were then so highly and deservedly appreciated, they can now be counted by the hundred, embracing among them every possible shade of colour from the faintest pink to the deepest crimson or scarlet, and from the purest white to the deepest purple, while the flowering season is protracted from February till the end of June—one variety following another without a pause during the whole period.

Rhododendrons are now more emphatically than ever the pride of European gardens, possessing as they do the most varied attractions to be found in any class of flowers, combined with the elegant habit and foliage of our finest evergreen shrubs, rendering them indispensable alike in the shrubbery, the flower-garden, and the conservatory—in each of these occupying a unique place, and never failing to elicit the highest admiration of all who have any appreciation of the symmetrical in form or the beautiful in colour.

This wonderful improvement has been effected chiefly through the crossing of the hardy late-flowering species with the early Indian sorts, particularly *arboreum*, itself too tender for open-air culture in this country; and it is to it that we are indebted for all the shades of scarlet, crimson, and pink, which are so much admired in our present race of hardy varieties. Among the first results of hybridising in this direction were the still well-known *Russellianum* and *Altaclarense* from Catawbiense varieties, *Smithii* from Ponticum, and *Noblicanum* from Caucasicum—all possessing in a greater or less degree the fine crimson flowers of their male parent. It was soon found, however, that though sufficiently hardy to stand the winter outside in favourable situations, these hybrids inherited much of its tree-habit and early-blooming peculiarity; they were shy in forming buds, or rather they required many years before they attained sufficient size for blooming; while in nine cases out of ten the flowers were blighted by late spring frosts before they were fully expanded, detracting immensely from their value for outdoor cultivation, and rendering it necessary for their safety that they should be potted in autumn, and placed in the conservatory after they were done flowering; and while they were, as they still are, extremely useful for this purpose, seeing that they can be had in all their glory during the winter and early spring months, it was obvious that much was still required to be done before bright-coloured Rhododendrons could be got to flower in May and June. Hybridisers naturally turned to the late species and their varieties; and these were again crossed, but with the crimson hybrids; and this carried on with every possible combination, and through a series of generations, has gradually developed that infinite variety of habit, form, and shade of colour with which our collections are now so much enriched. It would almost seem as if perfection itself had been attained, and a hybridiser may consider himself fortunate if, after having raised thousands of seedlings, he finds even one sufficiently distinct from, or even up to, the high standard of those already grown. The work, however, is still enthusiastically prosecuted, and from year to year novelties and improvements make their appearance.

Among the many novelties which, during the last quarter of a century, have been introduced into British gardens, none have attracted greater attention, or obtained a higher place in popular favour, than the magnificent species of Rhododendrons sent home in the spring of 1850 by Dr Hooker from the Sikkim Himalayas, and a few years later by Thomas J. Booth, Esq., from Assam and Bhotan.

The high expectations created by the glowing descriptions given by these gentlemen of their beauty in their native habitats—the great diversity among the individuals in habit of growth, from the tiny heath-like shrub up to the stately broad-leaved evergreen tree—the variety among them in foliage, form, and colour of flowers, and in many instances their remarkable dissimilarity in general appearance from any of the sorts hitherto known—have since then been fully realised; and they have proved invaluable acquisitions to the long list of these gorgeous-flowering shrubs already in cultivation.

It is scarcely possible to overrate the importance, in a horticultural point of view, of these brilliant discoveries, either as indoor decorative plants, for which many of them, from their noble foliage, delicate colours, and, in some cases, exquisite fragrance, are peculiarly adapted, or as the parents of a new and distinct race of hardy sorts. As most of the species, though hardy enough to survive our winters in the open air, in sheltered situations, bloom in winter or too early in spring, the great object to be aimed at is to infuse some of their peculiarities of foliage, colour, and, above all, fragrance, into the late-flowering varieties. Although, considering the length of time they have been in this country, comparatively little has been effected in this direction, from the difficulty of inducing some of the most desirable to cross with the hardy late-flowering American and European kinds, considerable progress has been made; and there are a number of hybrids which, while possessing many of the features of the Indian species, are at the same time late enough in flowering for outdoor cultivation. Much, however, remains yet to be done; and we can conceive of no more interesting and inviting field for the exercise of the hybridiser's skill, and in view of what has been already accomplished with *R. arboreum,* more likely to yield a rich return for his labours.

In crossing Rhododendrons with a view to obtain

hardy and late-flowering seedlings, we invariably choose the hardiest variety for the female parent, being satisfied from experience that the offspring inherit much more of the nature of the female than of the male in this respect; and when one of the sorts is tender, there is no species that we know of better adapted to form the basis of a hardy race than *R. Catawbiense*, which combines elegant foliage and showy flowers with an ability to bear any amount of frost that it can be subjected to in this country, along with a tendency to flower freely in a young state. It has been largely used by the most successful hybridisers, and its presence can easily be detected in some of our most useful and showy hardy hybrids.

R. maximum is also an excellent species for hybridising with the tender sorts, being hardy and late-flowering, and having fine compact trusses, the florets clear in colour and of a thick waxy texture, enabling them to resist hot sunny weather better than most sorts; but it has the drawback of being a shy bloomer, and is used with greater advantage after being first crossed with some of the free-blooming sorts. Some of the finest varieties have been obtained from its hybrids.

When early-forcing sorts are the objects in view, by far the best is *R. Caucasicum*, itself so early as frequently to expand its blush-coloured flowers in mid-winter in the open air. It is dwarf in habit, and a most prolific bloomer; and it has produced many most admirable varieties for such a purpose, including *Noblicanum* and *Noblicanum album*, first crosses by *arboreum* and *cinnamomeum*, both of which are invaluable for conservatory decoration from the middle of December till the end of February—a season when effective flowers are neither plentiful nor varied.

Notwithstanding their superlative claims upon the attention of all who have an interest in horticultural pursuits, and their undoubted value as decorative plants, whether arranged in masses in the flower-garden or shrubbery, or singly as specimens on lawns, it is somewhat remarkable that as yet Rhododendrons are neither so extensively nor so carefully cultivated as they deserve. In many establishments where they might be expected to be found occupying a prominent place, they are either absent altogether, or have the most inferior position assigned to them, and their sickly stunted appearance too often contrasts most unfavourably with the more robust and less fastidious shrubs with which they are associated.

This state of things, we are convinced, arises not so much from a want of appreciation of their merits, as from a popular mistake as to the difficulty of providing the necessary soil and conditions for their successful cultivation. Peat is not found in every garden, nor even in its immediate vicinity, and in many places the expense of procuring it in sufficient quantity is considered an insuperable barrier to the introduction of even a limited collection. That peat, or a combination in which the elements of which it is composed are largely present, is their natural soil, and that all known Rhododendrons thrive luxuriantly in peat, is undoubted; but that, at the same time, it is possible without it to create an artificial soil containing all the constituents which it supplies, really necessary for their growth and development, and this with materials to which most gardeners have ready access, has been again and again demonstrated.

On examining the root of a Rhododendron while in active growth, we find that it consists of what is commonly termed a ball or mass of roots, netting in a quantity of the soil in which it has been growing. Round the outside will be seen an innumerable quantity of short hair-like fibres, white and transparent, so extremely soft and brittle, that it is difficult to handle them without breaking some off; these are the young roots, and the only feeders by which the plant imbibes its food. If growing in peat, they are found pretty equally diffused over the ball; but if in mixed soil, they are invariably in greatest abundance on that side which is most in contact with any fragments of peat or other decomposed vegetable matter. Incapable from their extreme delicacy of penetrating stiff hard soil, and peculiarly susceptible of injury from dryness, particularly while in a state of activity, these fragile rootlets soon wither and die when so exposed, entailing a serious loss upon the plant, and that at a time when it requires all the assistance it can get to enable it to perfect its growth, and form flower-buds for the succeeding season. From these facts, as well as from experience of results, it seems obvious that a soil to be suitable for Rhododendrons must be soft and spongy in its texture, capable of retaining moisture, and possessed of a large percentage of vegetable matter.

Supplying all that is necessary for their sustenance, as far as soil is concerned, peat should invariably be used when attainable, in preference to any artificial compost; and when a choice can be had we prefer that which is found in bogs or heathy moors, as being less decomposed, more fibry, and consequently richer than that from higher and more exposed situations. The surface-turf only should be taken, cut not deeper than 6 inches, and chopped down with the spade sufficiently fine to allow the largest pieces to pass through a 3-inch sieve; after the addition of a moderate allowance of manure, which has been laid up at least twelve months—

say 1 ton to 6—with a similar quantity of clean sharp sand, the whole turned over and thoroughly incorporated, it may at once be transferred to the beds, and the planting commenced forthwith.

As we have already indicated, a limited supply of peat, or even its entire absence, need not deter any one from attempting the cultivation of Rhododendrons; the materials for forming an artificial compost which will adequately supply all their requirements, exist in a separate state, and may be found in abundance in every district in the country. Loamy turf from old pastures, cut just deep enough to include the fibre, of which it cannot have too much, with about one-half of its bulk of rotten leaves and old cow-dung, and more or less sand, according to the character of the loam, the whole mass chopped down with the spade, not too fine, and well mixed together, will form a compost which any Rhododendron will duly appreciate, and grow in with the greatest luxuriance. Charred garden refuse— such as prunings, weeds, and old tan-bark—forms a valuable supplement to such a compost, and may be used liberally, when it can be had, with great benefit to the plants. In no other form should these substances be introduced, as unless they are so thoroughly decomposed as to be scarcely distinguishable from fine mould, they are not only worthless but highly pernicious. This applies specially to old tan, which has sometimes been recommended. We have never seen the young roots working freely amongst it in any state; but very often when the decomposition was but partial, the ball was found to be covered with white fungus, and the plant in a sickly condition.

In connection with the making up of such composts, it may be noticed that, from some cause which we have never heard satisfactorily explained, Rhododendrons have the greatest repugnance to calcareous soil, and refuse to grow where lime or chalk is in immediate contact with the roots. Along with suitable soil, much depends for the successful cultivation of Rhododendrons upon the selection of a proper situation. Although the great majority of what are termed hardy sorts can bear any amount of frost they are ever subjected to in this country, they should invariably be planted in sheltered situations, as, particularly during the flowering and growing seasons, they are liable to damage from cold dry winds. When so exposed, their extremely delicate flowers become prematurely blighted, and the tender growths so much injured, as not only to prevent the formation of the buds, but often to cause a second growth, which is generally destroyed by autumn frosts.

Shade and drip from trees should also be avoided; for while the plants will grow vigorously provided they have sufficient moisture, they require a full exposure to the sun to enable them to ripen their wood sufficiently to withstand the winter's frost, and to form flower-buds. Even when they do flower in such circumstances, they develop neither their form nor colour to perfection. In preparing the beds for the compost, the best plan in most cases is to remove the old soil altogether, particularly if it is either stiff hard clay or worn out by long cropping. It is not requisite that this should be done deeper than one foot, as the tendency of Rhododendron roots is to spread out near the surface rather than downward, when the subsoil is stiff and retentive: nothing more is required than to fill in the compost; but in cases where it is loose and gravelly it is of importance, with a view to prevent its absorbing the surface moisture too quickly, that two or three inches of peat or leaf-mould should be forked in, and afterwards beat down firmly.

Although it is possible to transplant successfully all over the year, not even excepting the flowering and growing seasons, we would recommend that it should be confined to the autumn and winter months, beginning in October, by which time the buds are developed, and the young wood ripe enough to bear removal without injury. They should never be planted deeper than to allow the top of the ball being covered with about two inches of soil, which should be trod as firmly over and around it as possible.

In excessively dry summers a slight mulching with short grass, manure, or some similar material, will be found beneficial, by preventing evaporation and keeping the roots cool; while an occasional copious watering during the blooming and growing seasons will contribute largely to the vigour of the growth, and assist materially in the formation of large, sound flower-buds.

With a situation sheltered, yet fully exposed to the sun, a competent supply of suitable soil, a good amount of moisture, without wetness, and an occasional watering when exceptionally dry weather occurs during the growing season, Rhododendrons will grow vigorously enough for all practical purposes without being stimulated with manure in any form. Over-luxuriance is neither necessary nor desirable. On the contrary, when the primary object is abundance of bloom, it is rather prejudicial than otherwise, causing them to devote an undue amount of energy to the formation of mere wood, to the detriment of the flower-buds—and often inducing a second growth, which seldom ripens in time to escape damage from early frosts. Any tendency of this kind, whether arising, as is frequently the case, from excessive moisture at the root, or the soil being too rich, should be checked when the young shoots are sufficiently advanced to make it desirable that they should ripen and form their buds; and this may be effected by simply removing the soil from round the ball for a few days, or pruning the roots moderately with the spade.

All the varieties, however, delight in rich manures, and in cases in which the plants, by their weakly growths and sickly appearance, indicate a deficiency of the supply of nourishment, an inch or two of rotted manure, either forked in among the roots or laid on the surface as a top-dressing, with a slight covering of sand or light soil, in the course of the winter or early spring, will be found most beneficial; while the occasional application of a dose of liquid manure of medium strength, when the roots are in an active state, and even when the flowers are expanding, will work wonders in promoting their health and enabling them to make robust shoots and fresh well-developed leaves. We have seen a large collection, in which the plants had been for years in a most unsatisfactory condition—rarely flowering, and producing puny attenuated shoots—materially improved by such means. It may at the same time be observed that such a state of things will not be permanently remedied by mere stimulants, seeing that it is the result of either an exhausted or unsuitable soil; and the only effectual cure consists in lifting the plants and completely renewing the compost.

Considerable diversity of opinion has prevailed among cultivators as to which mode of propagation is best calculated to insure permanently handsome and healthy specimens. Some have contended that plants on their own roots, either obtained from seeds or layers, are superior to such as are grafted. We believe, however, that no universal rule can be laid down on the subject. All the species are easily raised, and make admirable plants from seed; but hybrid varieties will not thus reproduce exactly the same characters as the parents; while layering, except in the case of the few sorts that are naturally dwarf and branching from the roots, such as *Caucasicum* and *Noblicanum*, is at once a tedious and expensive process, necessarily limited in its application, and possessing few, if any, advantages over grafting. The plant so operated on is sacrificed, or at least hopelessly disfigured; and, after all, the product in young plants is at the most so trifling as to afford no compensation for the years of care necessary to make them presentable in the permanent beds among other specimens. There are, moreover, some of the richest-coloured and most attractive varieties which on their own roots have a rank, robust habit of growth, and consequent shyness in flowering, unfitting them for a place in general collections, but which are rendered all that could be desired by grafting.

The best and most commonly-used stocks for grafting are free-grown seedlings of the robust form of the common Ponticum, and the future wellbeing of the plants depends largely upon their being well selected. If they are weakly and stunted to begin with, failure and disappointment will be the sure result, and no amount of cultural skill will ever make them effective or creditable specimens. For the first two or three years after grafting, the stocks, if healthy and vigorous, have a tendency to throw up suckers from the root. These materially weaken the graft if allowed to remain, and should be carefully removed; and as it acquires strength to absorb all the sap the root can supply, they will gradually disappear. The finest specimens are produced from plants on single stems, entirely clear of branches for at least 6 inches above the root; when they are intended to stand singly in prominent situations, even 12 or 18 inches will not be found, after a few years' growth, too much. The lower branches soon bend down sufficiently to clothe the stem, while the head acquires a symmetry and uniformity of outline always pleasing, and which can never be attained by plants with an irregular mass of stems emanating from the root. In ordinary seasons most of the varieties set their seed freely—indeed, almost all the capsules will be found full; but unless wanted specially, it should never be allowed to ripen, as it entails a severe and unnecessary tax upon the strength of the plant; and the best course is, immediately after the blossom is decayed, to pinch the trusses off, which at that early stage may easily be done with the finger and thumb.

For forcing, to decorate the conservatory in winter, Rhododendrons occupy an important place, and one for which it would not be easy to find substitutes. A great many of the sorts are available for this purpose; and by a judicious selection, along with skilful management, a constant succession may be secured from December till the genial warmth of spring and early summer tempts those that are outside to expand their blooms. They should be potted as soon as possible after the flower-buds are fully formed, and placed in an open sunny situation out of doors until severe frost (to which they ought never to be exposed) necessitates their removal to more comfortable quarters. The earliest lot may be introduced into heat about the beginning or middle of December, according to the time they are wanted in flower, or the amount of heat that can be applied. From two to three weeks will generally suffice in an ordinary plant-stove temperature to expand the earliest varieties; while those that are later will require longer periods, according to their natural season of flowering when out of doors.

In potting, the balls should never be reduced more

than is absolutely necessary. It should always be borne in mind that every root cut off inflicts an injury upon the plant; and though they will expand their blooms after being considerably mutilated and cramped into unnaturally small pots, the evil effects of such treatment will be seen for years after.

While forcing, a moist atmosphere should be constantly maintained, the plants twice a-day syringed overhead, and the roots abundantly supplied with tepid water, with an occasional dose of weak liquid manure after the buds begin to swell. After the flowers are fully expanded, and not a day before, as they will be completely checked and make no further progress with a sudden change of temperature, they may be transferred to the conservatory, where the more they are shaded from the sun, the longer they will continue in perfection, adding to its attractions, and eliciting from even the most unimpressionable of its visitors the warmest expressions of admiration.

Forced plants, after their blooms are decayed, should be kept under glass till such time as they can be put outside without danger of suffering from frost, to which, after the heat they have been subjected to, they are peculiarly susceptible; and as they very rarely set a sufficient number of buds the first season to make them eligible for the following winter's work, they should be replanted in the borders as soon as the weather is mild enough to permit its being done with safety.

Since the first introduction of the Indian species, several no less useful and interesting forms have been added by other collectors from the same and similar localities, which, together with the varieties obtained from time to time by hybridising, constitute the now somewhat large group known as Greenhouse and Conservatory Rhododendrons.

Though the great majority of the species have been found able to survive our winters in the open air, and in one sense may be said to be hardy, they for the most part, as we have already remarked, start too early into bloom and growth; and, as a rule, the expanding flower-buds and young shoots are destroyed by spring frosts, and this even in the mildest seasons and most favourable situations. For the full development of their beauty, therefore, the shelter of a greenhouse is absolutely necessary; and they will richly repay their occupancy of the best place that can there be assigned to them, as they are undoubtedly the most interesting plants it can contain, especially during the spring months, either while in bloom or making their new growths, when they are singularly attractive and interesting.

The conditions necessary for the successful cultivation of greenhouse Rhododendrons are at once simple and easily supplied, and few plants are less exacting upon the care and attention of the cultivator.

In common with the outdoor members of the genus, they delight in a rich fibrous peat soil, which, with the addition of more or less sharp sand to make it sufficiently porous, alone should be used for their cultivation in pots.

In the preparation of the soil for potting, the turf should be chopped down to the requisite fineness with the spade, so as to retain all the fibre it contains; and as it must be pressed as firmly round the ball as possible, care should be taken that it be well aerated, and only used in a dryish mellow state. To insure perfect drainage, a thin layer of moss or rough peat should be put over the crocks before the finer soil is introduced.

The size of the shift must always be regulated by the habit and vigour of the individual plant, as well as by the state of the roots. In no circumstances is a large shift desirable; it is far better that it should be moderate, and repeated at such intervals as the growth of the plant and the spread of the roots render necessary. Apart from the importance, on the ground of convenience, of having them in as small pots as is consistent with their real wants, there is the possibility of the soil, from the necessity of frequent watering, becoming sour and sodden—a condition most prejudicial to their health, particularly in the case of such as are naturally slow-growing and weakly in their habit. In shifting, the ball should never be disturbed more than is necessary for the removal of loose soil, and on no account should the roots be torn off or mutilated. Of all the seasons of the year we prefer the spring for repotting, immediately after blooming, and when the plants are beginning to give indications of growth; the roots then at once take to the fresh soil, and the increased nutriment enables them to make rich luxuriant shoots, and consequently full and well-developed flower-buds.

During the growing season, the temperature of an ordinary greenhouse, except in the cases of a few of the more tender kinds, will be found amply sufficient; at this period they require an abundant supply of water at the roots, and a frequent sprinkling overhead. Such as have not been shifted, and have any appearance of being pot-bound, will be benefited by the application of weak liquid manure once or twice during the progress of their growth. As soon as the shoots begin to give indications of reaching maturity, water should be gradually withheld; and when the growth is complete, they should have no more water than what is barely necessary to keep them from flagging, and as soon as possible be removed to a cool situation out of doors, the great object being to prevent them from making second growths, to which most of the species have a strong tendency. If plunged to the brim of the pots in a sunny, sheltered situation, little artificial watering or attention will be necessary during the summer months, at the end of which they should be housed along with the other greenhouse plants.

The immense array of distinct species and varieties of Rhododendrons now in cultivation renders the selection of a moderate number, particularly for those unacquainted with them, a task of no little difficulty; and with the view of assisting such, we have arranged the following list of distinct and desirable varieties in groups and sections, according to their habits of growth, colours of flowers, time of flowering, and their uses in garden decoration. It is necessary to remark, however, that while all we have noted are really fine, they by no means embrace the whole of those that are worthy of cultivation; on the contrary, it would be easy to extend our list to a much greater length, and to name many others in each group equally deserving. It is therefore to be regarded rather as representative than exhaustive, from which a beginner may cull a good nucleus for the formation of a first-rate collection.

GROUP 1ST.—GREENHOUSE AND CONSERVATORY SPECIES
AND VARIETIES.

Section 1st.—Dwarf-growing, free-blooming sorts for greenhouse culture.

Ciliatum (Sikkim), a dwarf bushy species, forming a neat and effective pot-plant, blooming freely when only a few inches high; flowers, when newly expanded, of a blush colour, turning afterwards pure waxy white. It is hardy in mild districts; but as it flowers early, it is best treated as a greenhouse sort.

Countess of Haddington (hybrid) has a general resemblance to, but is much more robust in habit than, the preceding, of which it is a hybrid. It is a singularly beautiful, free-flowering variety, with blush-white flowers.

Fragrantissma (hybrid), of a dwarfer habit, but resembling *Edgeworthii*, of which it is a hybrid, with white fragrant flowers; the florets slightly pencilled with rose on the outside. It is a distinct and elegant plant.

Formosum (Gibsonii) grandiflorum, a well-known, free-blooming variety, with pure white flowers: one of the showiest of greenhouse sorts.

Henryanum (hybrid), a seedling from *Dalhousianum*, raised by the enthusiastic hybridiser, Mr Isaac Anderson Henry of Edinburgh. It is dwarfer in habit than the species; the flowers are pure white, deliciously scented, and very profuse. It is a grand acquisition to the list of greenhouse Rhododendrons.

Jasminiflorum, a beautiful species, with white tube-shaped flowers, arranged in clusters.

Javanicum (Java).—This species has beautiful tube-shaped, bright orange flowers, which it usually produces freely. It is of a compact habit of growth, and when properly managed, forms a fine specimen plant. It requires, during winter and spring, the temperature of an intermediate house, and to be kept in the greenhouse in summer.

Multiflorum (hybrid), a very dwarf, profuse-blooming variety, with white flowers resembling those of an Indian Azalea. They are produced at the axils of the leaves, and have, when expanded, a very pretty effect.

Princess Alexandra (hybrid), a dwarf variety, with a compact habit of growth. The flowers are long, tube-shaped, pure white, and the stamens delicate pink.

Princess Alice (hybrid), a dwarf, close-growing variety, with bell-shaped flowers, pure white, shaded on the outside with clear pink; they are sweet-scented.

Princess Helena (hybrid), a very beautiful variety, with long, tube-shaped flowers, produced in clusters; colour delicate pink, streaked with a darker shade. It blooms very freely, and forms a neat specimen plant.

Princess Royal (hybrid), resembling the last, but sufficiently distinct, the flowers being of a richly-shaded rose colour.

Virgatum (Bhotan).—This species has an extremely dwarf, bushy habit of growth; blooms freely when only a few inches high. The flowers are produced at the axils of the leaves, creamy white, shaded with rose at first, afterwards pure white.

Veitchianum (hybrid).—This beautiful variety has large, snow-white flowers, with a yellow blotch at the base, the margins prettily fringed. It is a distinct and very desirable plant.

Veitchianum lævigatum, resembling the preceding, but having the edges of the petals smooth instead of fringed.

Section 2d.—Species and varieties of tall and robust habits of growth, suitable for large conservatories.

Arboreum (Sikkim), the grandest of all the Rhododendrons. Flowers bright crimson or scarlet; its robust tree-like habit, however, renders its admission into glass structures of ordinary dimensions impossible; but where it can be accommodated, it is unexcelled.

Arboreum album, a variety of the last species, with white flowers.

Argenteum (Sikkim), a tall-growing species, with large heads of white flowers, somewhat shy in flowering, but much admired for its magnificent foliage, the leaves being from 6 inches to 1 foot long, and from 4 to 5 inches broad, silvery on under side.

Aucklandii (Sikkim), one of the finest of the Indian species; flowers snowy white, of a waxy texture, bell-shaped; the leaves from 6 inches to 1 foot long, and from 2 to 3 inches broad.

Blumeii (Java), a beautiful species, with delicate lemon-coloured flowers; blooms freely in a young state.

Dalhousia (Sikkim), a well-known and splendid species; flowers freely when of a moderate size; the flowers are white, 3 to 4 inches long, as much across the mouth, and fragrant.

Duchess of Buccleuch (hybrid), a seedling from *Edgeworthii*, with a general resemblance to that species; the flowers are large, waxy white, deliciously fragrant, and produced very profusely.

Edgeworthii, a magnificent species, with large, white, strongly and deliciously scented flowers; the leaves, for which alone it is worthy of cultivation, are prominently wrinkled above, and, along with the young shoots, clothed beneath with a dense woolly substance; the young leaves pure white, and changing as they become matured to a rich brown, giving the plant a curious and beautiful appearance.

Fortuneii (China), flowers of a delicate pinky-white colour, with a bright yellow throat, cup-shaped, from 3 to $3\frac{1}{2}$ inches in diameter; very sweet-scented.

Falconerii (Sikkim), a noble species, shy in blooming when young, but much appreciated for its noble foliage, the leaves measuring from 1 foot to 18 inches long and 7 inches broad, and beautifully ferruginous beneath. It is a grand object in a large conservatory.

Hookerii (Bhotan), allied to *Thomsonii*, to which it bears a strong resemblance. It is, however, quite distinct, and equally desirable for a conservatory. It was discovered by Mr Thomas J. Booth, and named in compliment to Dr Hooker. The flowers are bell-shaped, and of a very dark crimson colour. It is rather a shy flowerer while young or on its own roots, but very free when grafted.

Jenkinsii (Bhotan), a handsome, somewhat sparingly-branched species, bearing large, funnel-shaped, snowy-white flowers; it is a lovely conservatory plant.

Longifolium (Bhotan).—This is a grand species, with leaves above 1 foot long, and from 3 to 5 inches broad, strongly reticulated, and silvery beneath. The flowers, when they first expand, are of a delicate primrose colour, changing afterwards to a pure white.

Maddenii (Sikkim) resembles *Edgeworthii* in its large, finely-scented, white flowers, but in foliage quite distinct; an admirable pot-plant, with a bushy habit of growth, and though growing to a large size, flowers freely while comparatively small.

Nutallii (Bhotan).—This noble species, whether as regards foliage or flowers, is unsurpassed among conservatory plants; leaves 6 to 10 inches long by 3½ to 6 inches broad, beautifully reticulated; flowers trumpet-shaped, 4 to 5 inches long; colour white, with tint of rose red, and yellow at the base; delightfully fragrant. It is the largest-flowered Rhododendron known, and has a striking effect when in bloom. Being a tall-growing, sparingly-branched plant, it requires a large conservatory to develop its character to advantage.

Thibaudense (Bhotan), a remarkable and handsome species, different in general appearance, as well as in flowers, from any known Rhododendron except *Keysii*, to which it is allied. The flowers are tube-shaped, of a bright red colour, edged with yellowish green, with a translucent lustre, and produced in terminal trusses; a singularly interesting and beautiful plant.

Thomsonii (Sikkim).—This fine species has rich crimson flowers, which, when grafted, it produces freely, but requires to grow to a considerable size on its own roots before it begins to flower; it has pretty glaucous leaves.

Thomsonii hybrida (hybrid), resembling the parent in foliage and flowers, but blooms much more profusely. It is quite hardy enough for outdoor culture, but rather early for the majority of our winters. It is a first-rate variety for forcing.

Group 2d.—Early Species and Varieties suitable for Forcing.

Most, if not all, of the sorts of Rhododendrons in cultivation are available for forcing, and may be induced to flower considerably earlier than they would naturally out of doors; some, however, are better adapted for that purpose than others. Of such, the following are selected as the most showy and distinct. Those in the first section are very effective in mild springs in the open air, but so early that they can never be depended upon, a few degrees of frost being sufficient to destroy them. The others, though generally blooming in May, are also liable to have their flowers damaged, and require protection before and after they are expanded against the morning frosts, so frequent at that period of the season.

Section 1st.—Earliest-flowering sorts.

Albertus (hybrid).—This variety has light-pink flowers, changing in heat to an almost pure white; it is free-blooming, and very easily forced into flower.

Caucasicum (species), indigenous to high rocks on the

Caucasian Mountains. It has straw-coloured flowers, changing when under glass to an almost pure white. It is a very dwarf, compactly-branched plant, blooming year after year with the greatest profusion in mild winters in the open air, as early as December but generally in February or March. It requires only to be potted in the autumn, and kept in a cold frame or greenhouse, to insure its flowering in the middle of winter.

Caucasicum grandiflorum is a variety of the preceding, with flowers slightly darker, of a larger size, and with a more robust habit of growth; it is quite as early as the species.

Cinnamomeum (species).—This is an Indian species; flowers white, with black spots. It is a robust grower, but flowers freely when grafted.

Cinnamomeum hybridum; white flowers, very profuse, and of a dwarf bushy habit of growth.

Dauricum (species).—This is a deciduous species, indigenous to Siberia. The flowers, which are produced before the leaves expand in spring, are deep rosy purple, and generally very abundant. It is an admirable plant for winter work.

Dauricum atrovirens, deeper in colour, semi-evergreen, but otherwise very like the species, of which it is in all probability a seminal variety.

Diadem (hybrid).—This variety has clear pink flowers, and a habit of growth resembling *Caucasicum*, of which it is a hybrid.

Noblicanum (hybrid), flowers dark damask, in some varieties light pink; of a dwarf bushy habit; forces very easily. It is a well-known and popular hybrid variety, and not yet surpassed for early forcing.

Noblicanum album (hybrid), similar in habit to the last, but with snowy-white flowers, and equally early.

Præcox (hybrid), flowers rosy purple, resembling *Dauricum*, of which it is a hybrid, but more compact in habit. It is a remarkably free flowerer, and a superb plant for early forcing.

Præcox rubrum (hybrid) differs only from the preceding in the colour of its flowers, which are of a lighter shade.

Section 2d.—*Later species and varieties.*

Alta clerense (hybrid), flowers dark scarlet, with black spots ; rather shy in blooming on its own roots while young, but very free when grafted. It is a grand variety for the open air, but too early for the majority of seasons.

Alstrœmeroides (hybrid), flowers bright rose, spotted on all the petals.

Campanulatum (species).—An Indian species, with delicate blush flowers, changing to white ; blooms freely only when grafted, or when it has attained a large size.

Campanulatum hybridum (hybrid), flowers white with reddish spots, dwarfer in habit than the species ; grafted plants flower very profusely.

Caucasicum pictum (hybrid), flowers blush, or when forced, white, beautifully spotted with crimson. It has a neat, dwarf habit of growth, and is not only a first-rate forcing variety, but one of the finest for outdoor culture ; blooms about the last week of May.

Comet, an *arboreum* hybrid, with fiery crimson flowers ; very showy.

Cleopatra (hybrid), flowers blush, changing to white ; flowers very freely, and is easily forced.

Gloire de Gandavensis (hybrid).—A Continental hybrid, with white, finely-spotted flowers ; a magnificent plant when in bloom.

Guttatum colorans (hybrid), flowers snowy white, with prominent dark spots, equally valuable for forcing and outdoor culture. It is one of the most beautiful of the May-flowering hybrids.

Jackmannii (hybrid), flowers rosy purple, finely spotted, with a neat, compact habit of growth.

Jacksonii (hybrid), flowers light rose or pink with dark

spots; a remarkably free bloomer, and one of the finest early outdoor hybrids. It has a dwarf, compact habit of growth; blooms naturally about the end of May.

Lavinia (hybrid), flowers light rose, rather small, but very pretty.

Medora, flowers pink, very much spotted.

Prince Camille de Rohan (hybrid), flowers blush, with a rose eye; a remarkably abundant bloomer, with a compact habit of growth, and fine warm green foliage. It is a magnificent early variety for the open air, expanding from the middle to the end of May.

Princeps (hybrid), flowers brilliant scarlet, produced very freely. It has a bushy habit, and forces well.

Regalia (hybrid), flowers large dark crimson, and very early.

Russellianum (hybrid) resembling the preceding, but with flowers of a brighter colour. It has a robust habit of growth; the foliage is large, and of a fine deep green.

Smithii elegans (hybrid), flowers brilliant scarlet, rather small, but produced on grafted plants in great abundance.

Varium (hybrid), flowers deep rose, changing to blush. A distinct and pretty variety; free-flowering, and easily forced.

Venus (hybrid), flowers silvery blush or white when forced. It has a neat, dwarf, compact habit of growth, and in moderate heat flowers very early.

Vestitum coccineum (hybrid). — This variety has vivid scarlet flowers, with small compact trusses, and is one of the finest of its class.

Verschaffeltii (hybrid), flowers blush, with prominent black spots; an easily forced and remarkably beautiful Continental variety.

Group 3d.—Late-Flowering Rhododendrons.

Section 1st.—Varieties having white and light-coloured flowers.

Album elegans, blush, nearly white; very showy; flowers very freely; has a tall and erect habit of growth.

Album grandiflorum, white flowers, green eye; tall grower.

Amethystine, blush tipped with puce, with a bright yellow eye.

Athene, white, with a prominent yellow blotch.

Blanche, pure white, sienna spots.

Candidissimum, white, edged with the palest rose, prominent yellow spots.

Coriaceum, pure white; a very free flowerer; remarkably dwarf bushy habit, and fine foliage; very showy and valuable for a front row.

Chianoides, pure white, slightly marked with lemon spots; the truss very large.

Delicatum, French white, maroon spots.

Delicatissimum, clear white, delicately tinted with pink.

Evelyn, pure white; large, finely-shaped truss.

Empress Eugenie, waxy white, black spots; fine coriaceous foliage.

Fanny, clear white, orange spots; neat compact habit; fine foliage.

Gloriosum, white, large bold truss; very showy.

Hester, white, large compact truss.

Jean Stearn, white, with prominent crimson spots; shiny foliage.

Lady Godiva, white, yellow spots; large compact truss.

Luciferum, clear white, one of the best.

Lorenzo, white, bold yellow eye; very showy.

Minnie, white, very prominent blotch of chocolate spots; large conical truss, remains a long time in bloom, and is one of the most effective of the light-coloured varieties.

Mont Blanc, pure white, dwarf bushy habit, useful for a front row.

Mrs Hemans, white, yellow spots; an old but showy variety.

Mrs John Clatton, pure white, the florets of a thick waxy texture, the truss neat and compact, remains long in bloom; one of the finest whites extant.

Mrs Standish, pure white, yellow eye, large well-formed truss.

Mrs Tom Agnew, snow white, with pale lemon blotch, bushy habit.

Multimaculatum, white, red spots; the truss rather loose, and the florets small, but a very distinct and effective old variety.

Madame Miolan Carvalho, pure white, greenish-brown spots, compact neat truss, florets thick and waxy.

Papilionaceum, blush, nearly white, deep orange spots.

Ruth, blush white, prominent yellow spots.

Sultana, white, with reddish-brown spots; large, finely formed truss.

The Queen, very light mauve, changing to pure white; large compact truss, fine substance; a magnificent variety.

Zuleika, delicate blush, a distinct and pleasing variety.

Section 2d.—Varieties having lilac and purple flowers.

Amilcar, deep violet purple, strongly marked with dark spots.

Black-eyed Susan, light purple, heavy dark blotch.

Catawbiense, light purple, a well-known species, hardy and free-blooming.

Cyanum, purplish lilac, compact truss, very profuse.

Erestium, dark purple.

Everestianum, violet, pale-green spot; compact truss, the florets beautifully fringed; one of the best of the purples.

Fumosum, dark purple, strongly marked with dark spots.

Grace Darling, pale lilac, intense dark spots.

Grande, lilac, dark spots.

Leopardi, clear lilac, the whole of the bloom covered with red spots.

Lucidum, purplish lilac, brown spots.

Maculatum nigrum, dark purple; a superb free-flowering variety.

Maculatum purpureum, light purple, heavy black blotch.

Nereus, deep purple, dark spots.

Nero, rich bright purple, large truss.

Purpureum grandiflorum, dark purple.

Purpureum magnificum (Rembrandt), deep purple, fine bold truss.

Queen of Oude, violet shaded with white, dense black spots.

Sir Thomas Siebright, rich purple, bronze blotch.

Schiller, bluish purple, dark spots; very showy.

The Autocrat, fine purple, dark blotch.

Section 3d.—Varieties having rosy and crimson-purple flowers.

Alaric (Augustus), purplish crimson or plum.

Attila, deep claret purple.

Auguste Van Geert, light rosy purple, brown spots.

Blatteum (Sir Isaac Newton), claret crimson, large truss.

Currieanum, rosy lilac, spotted, compact truss.

Faust, rosy lilac.

Genseric, purplish crimson.

Gretry, rich rosy purple.

Iago, rosy purple, very dark spots.

Joseph Whitworth, dark purple lake, black spots, large truss.
Magnum bonum, rosy lilac, much spotted, very large truss.
Melanthauma, purplish crimson.
Murillo, rosy purple.
Ne plus ultra (*Londonense*), lilac purple, finely shaded.
Old Port, rich plum.
Omar Pacha, dark purplish lake.
Pardolotan, rosy lilac, finely spotted.
Prince Albert, rich lake, large shining foliage.
Queen Victoria, claret, a very free flowerer.
Sherwoodianum, rosy lilac, densely spotted.
Sir Colin Campbell, purplish rose, with a very dark blotch.
Stamfordianum, claret, black blotch.
Tamerlane, dark maroon, large truss.
The Maroon, chocolate, fine form.
Tippoo Sahib, very dark chocolate.
William Downing, dark puce, blotch of dark spots.

Section 4th.—*Bicolors, or varieties having the florets margined with a distinct colour from that of the throat.*

Alarm, white centre, margined with scarlet.
Baroness Lionel Rothschild, pale crimson centre, margined with intense scarlet crimson; a very beautiful and distinct variety.
Brayanum, rosy scarlet, light centre.
Bylsianum, white centre, margined with deep pink.
Claude, white centre, margined with lake.
Concessum, light centre, margined with bright rose.
Duchess of Sutherland, white, broad margin of rosy lilac.
Floretta, white, broad margin of cerise.
Fleur de Marie, white centre, margin rosy crimson.
Galbanum, light, rosy crimson margin.

Helen Waterer, pure white, margined with crimson.
Henry Bohn, pale, margined with rosy crimson.
John Spencer, rose, margined with deep pink.
Lady Grenville, white centre, margined with purple.
Limbatum, pale blush, edged with crimson.
Mrs John Walter, light centre, edged with brilliant crimson.
Mrs Layard, white centre, margined with crimson.
Mrs Thomas Brassey, white centre, margined with rosy purple.
Mrs John Penn, salmon, edged with lake.
Neige et Cerise, satiny white, margined with red.
Princess Mary of Cambridge, white centre, edged with rosy purple.
The Village Maid, light centre, bright pink edge, and yellow spots.

Section 5th.—Varieties having light-pink or rose-coloured flowers.

Albion, rose, fine form.
Amazon, delicate pink.
Bouquet de Flore, light rose, spotted.
Cato, rosy blush, finely spotted.
Congestum roseum, light rose, beautiful dark spots.
Distinction, rosy crimson, darker spots.
Duke of Norfolk, clear rose.
Erectum, deep pink.
Elfreda, light rose, beautifully spotted.
Flora, rosy blush.
Geranoides, rose, very dark spots.
Georgiana, light pink.
Giganteum, bright rose, large truss.
Gulnare, pale pink.
Ingomar, deep rose.
Jubar, light pink ; an exceedingly pretty variety.

Lady Armstrong, pale rose, much spotted.
Lady Easthope, clear rose, with dark spots.
Lady Eleanor Cathcart, clear rose, finely spotted.
Lady Emily Peel, bright rose, blotched with chocolate.
Lord John Russell, pale rose, dark blotch on upper petal.
Mooreianum, pale crimson, spotted.
Mammoth, pale rose, spotted.
Meritorium, light rose, yellow spots.
Metaphor, rose, fine form.
Mirandum, rose, faintly spotted.
Paxtonii, bright rose, chocolate spots.
Sir Charles Napier, rose, finely spotted.
Surprise, light rose, large truss.
The Gem, light pink.
Towardii, rose, spotted.
Prince Camille de Rohan, pink, dense crimson spots.

Section 6th.—Varieties having dark-crimson or scarlet-coloured flowers.

Alexander Adie, brilliant rosy scarlet, fine truss.
Atrosanguineum, intense blood-red.
Blandyanum, rosy crimson.
Brayanum, vivid scarlet.
Brennus, rich crimson lake, large truss.
Cabrera, clear rosy crimson.
Caractacus, rich crimson.
Cephalus, dark crimson, black spots.
Charles Dickens, dark scarlet, fine foliage.
Correggio, dark scarlet; very fine.
Comet, fiery crimson.
Comte Gomer, deep rosy crimson, compact truss.
Decorator, clear bright scarlet.
Dictator, dark crimson.
Duke of Cambridge, cerise, black spots.

Earl of Shannon, rich crimson.
Frederick Waterer, intense fiery crimson, large compact truss.
Francis Dickson, brilliant scarlet.
Gemmatum, bright scarlet.
General Canrobert, dark scarlet, cluster of dark spots.
Guido, deep crimson.
H. W. Sargent, crimson, very large trusses.
Hector, bright crimson.
Ignescens, bright scarlet.
James Bateman, clean rosy scarlet.
John Waterer, scarlet; fine substance.
John Gair, bright rosy crimson, large globular truss; very fine.
John Walter, rich crimson, neat truss.
Kate Waterer, clear rosy crimson, with yellow blotch.
Lord Brougham, crimson.
Lord Clyde, deep blood-colour.
Lord Granville, clear cerise.
Meridian, dark crimson, black blotch.
Michael Waterer, bright scarlet crimson.
Mrs Fitzgerald, bright rosy scarlet.
Mrs John Waterer, bright crimson, spotted; very fine.
Mrs Halford, salmon crimson, neat compact truss.
Narcissus, clear bright rosy scarlet, dark spots.
Neilsonii, very bright cherry-red.
Ornamentum, bright rosy scarlet.
Ornatum, dark scarlet.
Poussin, crimson.
President Van den Hecke, crimson, finely spotted; very fine.
Regificum, rosy scarlet.
Satanella, rosy red, fine.
Sidney Herbert, bright carmine, dark blotch.
Sir Robert Peel, brilliant rosy crimson.
Sir Henry Mildmay, bright rosy crimson.

Sun of Austerlitz, brightest scarlet.
The Bouncer, scarlet crimson, dark eye.
The Grand Arab (Vesuvius), brilliant crimson.
The Flamer, rosy scarlet.
The Grenadier, clear crimson, dark spots.
The Warrior, bright rosy scarlet.
Vandyke, bright crimson, compact truss.
William Cowper, bright scarlet.
William Austin, very bright crimson, large truss.

Section 7th.—*Varieties having double or semi-double flowers.*

Duc de Brabant, blush, with clear red spots, semi-double; has a bushy habit of growth.

Fastuosum plenum, lilac; an immense truss of large double flowers, which remain a long time in perfection. It is a most desirable variety, of very free growth, and flowers very profusely.

Hyacinthiflorum, purple, with neat compact trusses of small double flowers; an old but very fine variety.

Pyramidale plenum, white, with reddish spots, and a large pyramidal truss of semi-double flowers. It flowers very freely, and is very showy.

Section 8th.—*Species and varieties having dwarf bushy habits of growth, with small myrtle-like leaves, suitable for rockeries, small beds or borders, and edgings for clumps of the taller varieties.*

Arbutifolium, lilac; a close, bushy, dwarf shrub, with very dark-green foliage.

Chamæcistus, pink, purple-shaded. This is a species indigenous to the Alps of Austria, and other high mountain-ranges in the colder countries of Europe. It is a tiny, heath-like shrub of about 6 inches high, very pretty when in flower, and well adapted for a rockery or small garden bed. It should be planted in gritty peat, with the ground well drained.

Ferrugineum, rose; a low-growing species from the high ranges of the Alps; is a well-known and beautiful little plant, with small leaves, green on the upper and rusty on the under surface.

Ferrugineum album, differs only from the species in having white flowers.

Gemmiferum, pink; a very beautiful hybrid.

Govenianum, pale pink, delicately scented; a pretty plant.

Hirsutum, deep rose, resembling *Ferrugineum*, but dwarfer in habit; the leaves covered with small hairs. It is also indigenous to the Alps, and, associated with that species, is the last ligneous vegetation before the line of perpetual snow.

Hirsutum variegatum, dwarfer in habit than the species, and with its leaves more or less covered with a golden variegation.

Myrtifolium, lilac, with fine light-green foliage, and very bushy.

Odoratum, pale rose, sweetly scented; an exquisitely pretty plant; should be planted in a sheltered situation, and is even worthy of greenhouse culture.

Waterer's hybrid, rosy lilac; a very free-growing variety.

Wilsonianum, rosy lilac, fine warm green foliage; fine bushy habit.

Section 9th.—Species and varieties suitable for game-covers, and for planting in woods for undergrowth.

This section is composed of *ponticum, catawbiense,* and the almost innumerable varieties which have resulted from hybridising. *Ponticum* itself is now planted by the thousand for game-cover, and is rarely, if ever, even in the most severe winters, injured by hares or rabbits. The many hybrids of which it is the parent enjoy an equal immunity from the ravages of these animals, and can all be recommended for any situation where they abound. Many of the varieties, though lacking that fine form of truss and delicacy of colour so much valued among the choicer varieties, are nevertheless very showy, and produce beautiful effects on the margins of woodland drives, banks of rivers, and park shrubberies. It is needless to say that both species are perfectly hardy, and that they thrive well in any soil rich in decayed vegetable matter. These varieties, with few exceptions, are sold in the nurseries as unnamed seedlings, and possess colours of all the various shades of rose, white, and purple.

Catawbiense.—A bushy species of from 10 to 12 feet high, indigenous to North America; very abundant in Virginia and Carolina, particularly near the source of the Catawba river. The flowers are rosy lilac, varying in the different varieties to blush and white.

Ponticum.—This species is found wild in Armenia and several other parts of Asia Minor, forming a large bush of from 10 to 15 feet in height. The flowers are of various shades of purple, sometimes nearly white.

AZALEA (THE AZALEA).

Possessing many features in common with the Rhododendrons, to which they are so closely allied that, with the single exception of our native species *procumbens*, which was, we think, properly referred to the genus *Chamæledon*, the older botanists classed them in that genus, the Azaleas are beyond all question in the front rank among hardy flowering shrubs.

Though less vigorous in their habits of growth, and lacking that imposing grandeur so much admired in the Rhododendrons, they recommend themselves by their remarkably profuse blooming qualities, and by the exquisite richness and variety of the colours of the flowers, embracing as they do all the shades of crimson, rose, pink, orange, yellow, and white; and these in the several sorts, blotched and striped in almost innumerable combinations, give an interest and charm to the American garden in May and early June which is at once peculiar and striking, and which must be seen to be adequately realised.

Nearly all the really hardy Azaleas are deciduous, and, with the exception of *pontica*, which was sent home from the Levant about the end of the last century, the few species that have formed the parents of the great majority of the now numerous varieties are natives of North America, and though introduced into Europe at intervals between the years 1734 and 1818, little was done for a long time by way of producing new varieties by hybridisation —the variations being chiefly sports, or in some instances natural hybrids, the result of the different sorts being grown together. About thirty years ago, however, some of the Continental cultivators took up the matter systematically, and the brilliant varieties known as Ghent Azaleas, a term now popularly applied to all the hybrids, were the fruits of their labours. This work of improvement has since been steadily carried on, both on the Continent and in this country; and in later years, the Chinese *sinensis*, a tender deciduous species, has been judiciously intermixed with the hardy sorts, the result being seen in the increased size, improved form, and clearer colours of the flowers of the newer varieties. So far, indeed, have the original American species — such as *calendulacea, viscosa, nudiflora,* and *speciosa* — been eclipsed and superseded by their progeny, that they are now almost out of cultivation in their normal state.

The oriental species are for the most part too tender for outdoor culture in Britain. Though one or two of those introduced a few years ago have been found equal to our climate in the open air—such as *obtusa* and *amœna*—they bloom and start too early into growth to escape damage from spring frosts, precluding their being so extensively planted in the open air, and from receiving that prominence to which, but for that unfortunate drawback, their beauty and free-blooming qualities so richly entitle them.

Of these species, *amœna* is perhaps the hardiest and most showy, producing its rich rosy-crimson blossoms in April, and forming a neat dwarf round bush, densely furnished with small dark-green leaves, rendering it, apart from its flowers, a useful marginal evergreen or semi-evergreen shrub for clumps or beds of plants of taller growth, and well worthy of a place in the American garden, even though it should sometimes require a little attention during its flowering season in the way of protection at night from frost. It is worthy of notice that it is a superb conservatory plant, and may be forced into flowering at Christmas with great facility, and if moderately shaded remains a long time in perfection; for this purpose they are either grown in pots and plunged in a sunny situation during summer in the open air, or lifted from the ground and potted early in November, and placed at once under glass, and introduced into the forcing-house from time to time as they are wanted in flower.

The arrangement and distribution of Azaleas in

the American grounds must always depend upon the taste and convenience of the cultivator. Under ordinary circumstances, they require neither artificial watering nor more shelter than is usually afforded to other hardy shrubs. They are found growing naturally in dry situations, and prefer gritty fibry peat; and though, in common with Rhododendrons, they require a good amount of moisture while making their growth, which may be supplied them with the greatest advantage when the weather is exceptionally dry during that season, they will not thrive in a low swampy position where the soil is saturated with water during winter; in such a situation their roots soon decay, and, as a necessary consequence, the plants become sickly and gradually die. In preparing ground for their reception, therefore, care should be taken to have it sufficiently drained to prevent the possibility of stagnation; and in the case of retentive clay soils, it is a good plan to raise the beds considerably above the surrounding level. When a choice can be had, preference should be given to a north-west aspect, as they are there protected from the full glare of the sun, and less apt to suffer from continued drought, while the partial shade keeps the flowers longer in perfection. They may be transplanted any time during the autumn, after they shed their leaves, till they begin to show signs of activity in early spring; in this operation the roots should be carefully preserved, and the ball as little mutilated as possible.

Azaleas being so distinct in general appearance from the rest of the American plants, give a pleasing variety and appear to great advantage when associated with Rhododendrons and other evergreens in mixed borders or beds; or when, as is frequently done, grouped in masses by themselves, they have a magnificent effect—their elegant foliage, of a fine warm green in spring and summer, changing to a bright red or yellow tint in autumn—and richly coloured, in many cases fragrant flowers, are produced in profusion, amply compensating for that naked appearance in winter which is sometimes urged as an objection to that mode of planting.

In the following list are a few of the finest varieties in cultivation, all of them thoroughly hardy, profuse in flowering, and most effective in colour:—

Admiral de Ruyter, dark red or scarlet; very large and showy.
Alba flavescens, yellowish white.
Alba flavescens rosea, yellowish white shaded with rose.
Ardens, dark red, small flowers, but very profuse.
Atrorubens nova, dark red.
Augustissima, orange, red, and yellow.
Aurantia major, orange.
Aurantia speciosa, pale orange.
Aurantiaca cuprea, orange and copper stripes.

AZALEA.

Baronne G. Pycke, yellow shaded with pink.
Beauty of Flanders, pale yellow and salmon.
Bicolor grandiflora, yellow and red.
Belle rosette, light sulphur and rose.
Calendulacea coccinea, orange and scarlet.
    „   *ignea,* yellow and bright red.
    „   *elegans,* yellow and pink.
    „   *eximea,* reddish orange.
    „   *crocea,* orange and red.
    „   *triumphans,* bright yellow and red.
    „   *punicea rosea,* yellow and rose.
Carnea, pink and sulphur.
Cruenta, sulphur and scarlet.
Cuprea eximea, copper and yellow.
    „   *grandiflora,* copper and sulphur; large flower.
Coccinea, dark scarlet, small flowers, but very abundant.
    „   *major,* dark scarlet, larger flowers.
    „   *maxima,* bright scarlet.
    „   *speciosa,* orange scarlet.
Decorata, delicate pink.
Delicata nova, pink and yellow.
    „   *rosea,* rose and yellow.
Duchesse d'Orleans, orange crimson.
Etendard, rich scarlet.
Elegans, light yellow, striped with red.
    „   *Mortierii,* sulphur and red.
Fulgens, orange and scarlet.
Gloriosa, light orange.
Géant des Batailles, deep crimson.
Genio Mortierii, orange and scarlet.
Gloria mundi, bright orange scarlet.
    „   *patriæ,* rosy pink.
    „   *triumphans,* yellow shaded with pink.
Ignescens, orange crimson.
Imperatrice, orange, yellow, and pink.
Ignea nova, light orange crimson.

Incarnata, pale flesh-colour.
Leopold Premier, sulphur, red, and orange.
Lutescens grandiflora, bright yellow, large flower.
Lateritia striata, white striped with pink.
Lineata, pink and yellow.
Mutabilis, whitish shaded with pink.
Magnificans rosea, pink.
Ne plus ultra, orange scarlet.
Odorata albicans, white ; very sweet scented.
    „   *pallida*, blush white.
Oscar I., light sulphur, shaded pink.
Pontica, yellow ; a well-known species.
    „   *alba*, blush or creamy white.
    „   *flammea*, yellow and red.
    „   *multiflora pallida*, pale sulphur.
    „   *tricolor*, orange, sulphur, and red.
    „   *Victoria modeste*, white, shaded reddish pink.
Præstantissima, dark scarlet.
Princesse d'Orange, salmon pink.
Rosea formosissima, pink or light rose.
Saltatoria, light yellow and red.
Sanguinea, deep crimson.
Superbissima, dark orange, large flower.
Versicolor, pink and creamy white.
Viola odora, orange and red ; very sweet scented.
Van Houttei, a splendid double-flowered variety ; orange, scarlet, and yellow.
Viscocephala, blush white, finely scented.
Venusta, deep orange and pink.
Viscosa floribunda, small white flowers deliciously scented.
Zenobia, orange and crimson.

ANDROMEDA (THE ANDROMEDA).

In this fine genus we have a rare combination of some of the qualities which are most desirable in hardy flowering shrubs. The neat habits of growth, elegant foliage, and showy flowers of the various species have long been appreciated, and have secured for them a prominent place in most collections of peat-soil plants. They are widely distributed over the colder regions of Europe, Asia, and America; and it is said that Linnæus, struck with the graceful beauty of their flowers as contrasted with the dreary wastes, popularly believed to be haunted by supernatural beings, which form their natural habitats, gave them the classical name Andromeda, in allusion to the old fable of the beautiful Ethiopian princess who was chained to a rock and exposed to the attacks of a sea-monster.

Most of the species are thoroughly hardy in this country, forming dwarf bushes, densely clothed with leaves—with very few exceptions evergreen—and producing in early spring and summer their lovely, wax-like, bell-shaped blossoms with the greatest profusion, uninjured by the severest frosts.

As regards their culture and general management, little more need be said than that, along with a moderate allowance of peaty soil, they should have a larger amount of root moisture than most other

American plants. They luxuriate in a marshy, swampy situation, in which Rhododendrons and Azaleas could not exist for any length of time; it is therefore important to keep this peculiarity in view when it is intended to plant them in beds or clumps by themselves—a mode of planting which, though not very generally adopted, is nevertheless very effective, from the great diversity in foliage, heights, and general appearance of the plants.

This preference for a damp situation need not, however, deter any one from associating them in mixed borders with the other peat-soil shrubs; they adapt themselves to such circumstances with the greatest facility, though their vigour will be promoted, and they will flower much more freely, if supplied with water when the situation is naturally dry, or in cases of long droughts, especially during the growing season.

All the hardy species being interesting and well worthy of cultivation, it is somewhat difficult to make a selection of what are usually termed the most desirable varieties; and while the following may be regarded as really fine and distinct, we would recommend those who intend planting, and who have the means and the necessary accommodation, to grow as many of the sorts as they can procure, fully assured that no more ornamental and pleasing hardy flowering-shrub can be introduced into a garden or pleasure-ground than an Andromeda, under whatever name it may be known:—

A. axillaris (*the Axil-flowered Andromeda*).—This is a low-growing evergreen shrub of from 1 to 2 feet high, indigenous to high elevations in various parts of North America, from whence it was first introduced to our gardens in 1765. The flowers are white, produced at the axils of the leaves in racemes, and are generally in perfection in May or early in June. The leaves with which it is abundantly furnished are slightly glabrous, and of an oblong or sharp-pointed oval form. It is a neat little plant, very distinct in its appearance, and makes a fine margin to a clump of the larger shrubs.

A. angustifolia (*the Narrow-leaved Andromeda*), an evergreen species indigenous to swamps in Carolina, Georgia, and other parts of North America, of a dwarf bushy habit of growth, rarely found higher than about 2 feet, and first cultivated in British gardens in 1748. It is an interesting little plant, of a neat habit of growth, and useful for small beds or edgings. The leaves are small, linear-lanceolate in form, green above and rusty beneath. The flowers are white, small, and not very conspicuous, though very pretty; they are usually in perfection in May or June.

A. calyculata (*the Large-calyxed Andromeda*).—A native of North America, found abundantly in bogs and swamps over a wide area from Canada to Virginia, as also in Siberia and other countries in Northern Europe. It forms a close evergreen bush of about 2 feet high, producing its pretty white flowers very freely in April and May. The leaves are elliptic-oblong, very small, deep green above and rusty beneath. It was introduced in 1748.

Var. *latifolia* is equally as interesting as the species, from which it differs in having broader leaves, larger flowers, and a more robust habit of growth.

A. floribunda (*the Many-flowered Andromeda*) from Georgia and other mountainous districts in North America, forming a broad compact bush rarely exceeding 4 feet in height; introduced in 1812. It is unquestionably the finest of the genus, and one of the most showy and beautiful of our hardy evergreens. The leaves are of a deep green colour, ovate-oblong in form, and very abundant. It produces its snow-white, waxy, bell-shaped flowers year after year with the greatest certainty, and with remarkable profusion, beginning early in February and continuing till April, and is so hardy that it is rarely affected even in the slightest degree by our severest frosts. A bed of this superb plant, margined with the early-flowering *Erica herbacea carnea*, is a sight in the spring months not easily forgotten. Small plants, potted in autumn, have a fine effect in the conservatory in winter.

A. polifolia (*Moorwort*), found wild in several of the moorland districts in England and lowlands of Scotland, in similar situations in some of the northern countries of Europe, and over a very extended area in North America, is a peculiarly interesting, very dwarf evergreen species, seldom growing higher than about a foot. The flowers, which are of a delicate rose colour, are produced abundantly in May and June. The leaves are small, very numerous, oblong, green above and glaucous beneath. It is an elegant plant for margins of beds of American plants. Of this species the following distinct varieties are favourites with cultivators:—

Var. *angustifolia* has much narrower leaves than the parent, but is similar in other respects.

Var. *major* is much more robust in habit of growth, and has larger leaves, than the parent.

Var. *rubra*—this variety has red instead of pink flowers.

A. pulverulenta (*the Powdery Andromeda*).—This, by some botanists classed as a variety of *speciosa*, is a deciduous form indigenous to North America, inhabiting the swamps of North Carolina, from whence it was first introduced in 1800. It is a very handsome bushy shrub of about 3 feet high.

The leaves and stem are thickly covered with a white powdery matter, which gives them a novel appearance, and renders the plant valuable as a contrast to the dark green of the other species. The flowers, which are in perfection in June, are pure white, large, and showy, and disposed in racemes.

Var. *pulverulentissima* partakes much of the character of the parent, and only differs in having a more robust habit, and in the leaves being more thickly covered with powder.

A. rosmarinifolia (*the Rosemary-leaved Andromeda*).—This is a dwarf bushy species, indigenous to the bleakest districts of Newfoundland and Labrador, introduced about 1790. It is one of the prettiest of the species, forming a dense bush of about a foot in height. The flowers, which are of a delicate pink colour, are in perfection early in June; the leaves are linear-lanceolate, slightly convex, and white beneath. This, though a common species, is a perfect gem, making a fine marginal plant, specially attractive when in early summer it is in flower, or making its young growths.

A. tetragonia (*the Four-cornered-branched Andromeda*), a native of Lapland, Siberia, and the coldest regions of North America, introduced in 1810, is a small heath-like shrub, of a prostrate habit of growth, rarely seen rising above 6 inches from the ground. The leaves are densely imbricated and adpressed in four rows round the stem; the flowers, which are pure white, are in perfection in March or April, but so small and inconspicuous that they give very little character to the plant. It is, nevertheless, a distinct and beautiful form, interesting for an edging—or for a rockery, if planted in the dampest part, and in a north aspect.

ARCTOSTAPHYLOS (THE BEARBERRY).

The few species of which this interesting genus is composed were formerly referred to *Arbutus*, and though by no means strikingly showy either in foliage or flowers, are yet sufficiently distinct for admission to even a small American garden.

They are natives of America and the colder countries of Europe, inhabiting dry rocky situations, and are valuable in our gardens for clothing rockeries and dry sterile banks on which it is difficult to induce other shrubby plants to grow. In garden cultivation it is necessary to raise the beds sufficiently above the surrounding surface to insure thorough drainage, so that the plants may ripen their wood before winter, as in damp or deep rich soils they have a tendency to continue growing till late in autumn, and to suffer damage from frost.

All the sorts flower freely, and produce abundance of berries, which are very pretty in winter. They are more or less procumbent in their habits of growth, and resemble in general appearance some of the well-known species of *Cotoneaster*.

A. alpina (*the Alpine Bearberry*), indigenous to all the countries of Northern Europe, several of the highest mountains in the Highlands of Scotland, and over a wide area in Canada, forming a trailing deciduous shrub, with obovate serrated leaves, of a deep green in summer, assuming as they decay in autumn a bright red tint. The flowers are of a pale rose colour, in some cases nearly white; they expand in May or June, and are followed by black berries ripe in September. It is a curious little plant, well worthy of a place on a rockery.

A. uva ursi (*the Common Bearberry*) has a similar geographical distribution to the preceding, and is also found in many mountainous districts throughout Britain. It is an evergreen, with long, robust, trailing stems, clothed very thickly with rigid, obovate, entire leaves. The flowers are pale red, produced in crowded racemes at the points of the branches, and are usually in perfection early in June. The berries are bright red, and ripe in September. It is a fine ornamental plant for a rockery or any dry situation, while its free growth and thorough hardiness commend it to the attention of those interested in the formation of game-cover; birds of all kinds eat its berries eagerly, and we are convinced that it might be planted extensively for that purpose on bare waste moors or hillsides, with the greatest success.

BRYANTHUS ERECTUS.

The tiny evergreen known to cultivators by this name is a hybrid obtained many years ago by the late Mr James Cunningham, of the Comely Bank Nurseries, Edinburgh, the parents being *Menziesia cærulea* and *Rhododendron chamæcistus*. The impropriety of referring it to that genus is therefore obvious, its true place being among the *Menziesias*, to which it is so nearly allied, and with which it has so many characters in common.

By whatever name it may be called, however, it is undoubtedly one of the most beautiful of our dwarf peat-soil shrubs, and ought to have a prominent place wherever such are grown. It forms a neat compact bush of about 6 inches high, and flowers profusely about June, the whole surface of the plant being covered with its large, bright pink, Kalmia-like flowers, which, being of a thick waxy substance, remain a long time in perfection. Like its parents, it is perfectly hardy, and though preferring a shady situation, grows and flowers freely in the ordinary peat-beds, when well drained. It is a superb rockwork plant, and invaluable for edging beds of the dwarfer American plants. It grows best in sandy peat, which in planting should be firmly pressed round the ball; and its vigour is promoted in dry seasons by a good soaking of water while making its growth.

CLETHRA (THE CLETHRA).

This genus, of which there are a considerable number of species known to botanists, is perhaps best known through its representative *arborea*, an old and still popular greenhouse evergreen of great beauty, but unfortunately too tender for our winters out of doors. There are, however, several fine hardy species, two of which we feel cannot be too highly recommended for admission to a collection of American shrubs: both are deciduous, and much valued in our gardens for their showy foliage, and for their pretty, conspicuous flowers.

All the species require a rich peaty soil, and to be planted in a dampish shady situation; and though both of those we describe are hardy, they always thrive best in a situation protected from high winds.

C. alnifolia (*the Alder-leaved Clethra*), a native of North America, and particularly abundant in the swamps of Carolina and Virginia, growing from 3 to 4 feet high, and much valued in our gardens as a distinct-looking ornamental shrub: it was introduced in 1731. The leaves are cuneate-obovate, acute, serrated, glabrous, and of a bright green colour; the flowers are white, produced in racemes at the points of the branches, and usually expand in July or early in September.

C. tomentosa (*the Downy Clethra*), is found wild in similar situations in Carolina and Virginia as *alnifolia*, forming a compact bush of about 4 feet high. It was introduced at the same time as that species and has proved equally hardy and valuable as an ornamental shrub. The leaves are of a cuneate-obovate form, serrated at top, dark green, and clothed on the under surface with minute white hairs. The flowers are white, produced in racemes at the points of the branches, and expand from August till October. It is a very handsome and desirable plant.

COMPTONIA ASPLENIFOLIA (THE FERN-LEAVED COMPTONIA.)

This, the only representative of the genus in cultivation, is a deciduous shrub of about 4 feet in height, belonging to the natural order *Myricaceæ*. It is found wild in peaty woods in the colder parts of Canada and the United States, from whence it was first sent to this country in 1714, and named in honour of Henry Compton, Bishop of London, who was celebrated as an enthusiastic botanist and horticulturist.

It is here a compact bush, with numerous short wiry branches, well clothed with oblong-linear leaves deeply cut on each side into rounded lobes, and resembling miniature fronds of the well-known fern *Ceterach officinarum*. The bark and foliage are thickly covered with minute resinous dots, which, when bruised in the slightest degree, emit a very pleasing aromatic fragrance. The flowers are of a rusty-brown colour, produced in small catkins from the sides of the branches, and generally in perfection in April. Though, like its near allies the *Myricas*, it has little, as far as flowers are concerned, to commend it to the favourable consideration of planters: it is nevertheless a very pretty shrub, interesting from its neat foliage and sweet fragrance, and valuable as a contrast in mixed shrubbery borders, from the peculiar brownish-green tint of the young leaves, caused by their being covered with down. It is only necessary to add, that notwithstanding the long period it has been cultivated in our gardens, it is still very rarely met with—a fact the more remarkable when we consider that it is quite hardy, and of free growth in the ordinary peat-bed if planted in a shady situation.

CHAMÆLEDON PROCUMBENS (THE ALPINE AZALEA).

This interesting little plant is also referred to the genus *Loiseluria*, but more commonly to *Azalea*, and is best known as *A. procumbens*. It is so very different, however, from any of the other cultivated forms of that genus, that we adopt *Chamæledon* as not only most distinctive, but as in all probability the most scientifically correct.

It is a procumbent evergreen shrub, indigenous to Alpine regions over a wide area in Europe and North America, and occurs in great abundance on most of the mountains of the Scottish Highlands, particularly on the Cairngorm range, growing always in greatest luxuriance in dry sandy peat among grass and moss, and producing its lovely pink bell-shaped blossoms in April and May. In general appearance it resembles the wild thyme, forming a close, carpet-like bush, never rising above a few inches from the ground: the slender wiry branches are amply clothed with small elliptic leaves, of a bright green colour. The Rev. Hugh Macmillan, in his pleasing and instructive book, 'Holidays in High Lands,' in describing this plant as seen in its native habitats, says: " Close beside the bridle-path which winds over the heights from Glenisla to Braemar, an immense quantity of the

Highland Azalea grows among the shrubby tufts of the Crowberry, and when in the full beauty of its crimson bloom, it is a sight which many besides the botanist would go far to see. It is the only plant on the Highland mountains that reminds us of the Rhododendrons which form the floral glory of the Swiss Alps, and especially of the Sikkim Himalayas."

As it grows freely in dry situations, when planted in sandy gritty peat, it is a charming edging-plant in the American garden, and is well adapted for rockeries, where, if supplied with the proper soil, it forms a most attractive little specimen, flowering year after year with the greatest profusion.

DAPHNE (THE DAPHNE).

This is a genus of handsome, low-growing, evergreen and deciduous shrubs, said to be named in allusion to the nymph Daphne of the old Greek fable, who was changed into a Laurel—some of the species having a resemblance to these plants.

They are found widely distributed over Europe and Asia, chiefly, however, in the temperate and colder regions, and consequently a large proportion of them are hardy enough for outdoor cultivation in Britain. The flowers of most of the sorts are delightfully fragrant.

Though all the species delight in a peaty soil, and are sometimes associated with the American shrubs with the happiest results, they are not usually regarded as part of the group, seeing that, with two or three exceptions, they are equally at home in rich loam or ordinary garden-soil; and while we confine ourselves to a description of these exceptions, we would at the same time remark that, where space permits, all the species should have a place assigned them in the American garden.

D. collina (*the Italian Daphne*).—This is a free-flowering evergreen shrub, found naturally on low hills and banks of rivers in the south of Italy, forming a neat, densely-branched bush, rarely exceeding 3 feet in height. It has been cultivated in Britain since 1752.

In most districts it is quite hardy, if planted in a moderately sheltered situation, and is sometimes seen in great perfection trained on walls. The leaves are obovate, bright glossy-green above and downy beneath. The pretty pink flowers begin to expand, when the season is mild, early in spring, and come out in succession till June; they are produced in terminal clusters, and are sweet-scented.

Var. *neapolitana*.—This fine variety is said to be a sport from the species, from which it is quite distinct, having lighter-coloured flowers, and leaves smooth on both sides. It was introduced from Naples in 1822.

Var. *dauphine*, also called *hybrida*, is another beautiful variety of hybrid origin. The flowers are bright red.

Var. *fioniana*, another hybrid, resembling, but quite distinct from, the parent. It forms a neat little bush, and flowers very profusely.

D. cneorum (*the Garland Flower*) is a procumbent evergreen indigenous to mountains in Switzerland, Hungary, Germany, and France, and has been cultivated in Britain since 1752. It is quite hardy here, and forms a dense bush, rarely rising higher from the ground than about a foot. The leaves are small, lanceolate, sharply pointed, and of a light shiny green. It flowers about the beginning of May, and in mild, damp seasons, for the second time in autumn, every twig producing its terminal cluster of delicately-scented, rosy-pink blossoms. The berries are pure white, but rarely, if ever, come to perfection in this country. For margins of clumps of the more robust shrubs, or for planting in a small bed by itself, no plant is more effective; and well does it deserve the best place and the kindliest treatment that can be given it, as there is not a lovelier gem among hardy shrubs. A deep sandy peat, and a sunny but not too dry situation, suit it admirably; and an occasional watering in very dry summers will be found to promote its vigour, and to induce it to flower in autumn.

The following are distinct and very desirable varieties, and, with the species, deserve a place in the choicest collections of ornamental shrubs.

Var. *variegata.* — This variety has its leaves narrowly margined with a straw-coloured variegation, but is otherwise identical with the parent, with which it contrasts beautifully in beds. It is a pretty little plant, and worthy of extensive cultivation.

Var. *major* has a more robust habit of growth, broader leaves, larger flowers, and blooms earlier than the species. It is a charming spring bedding-plant, and is sometimes planted with good effect in centres of beds of the dwarf form. Like the parent and many of the other dwarf free-flowering Daphnes, it is a fine plant for forcing for the early spring decoration of the greenhouse or conservatory, for which purpose they should either be grown in pots, but fully exposed to the sun in the open air in summer, and only housed on the approach of winter, or lifted from the borders in November, and placed in cold frames, to remain for a few weeks so that it may to some extent be established before being introduced into heat. We prefer, however, to grow the plants intended for forcing in pots, and to give them similar treatment to that we have already recommended for greenhouse Rhododendrons. This, we may add, applies not only to the hardy sorts, but to *D. indica* and the other tender or half-hardy species.

DIRCA PALUSTRIS (THE LEATHER-WOOD).

A low-growing deciduous shrub, seldom growing higher than from 3 to 4 feet, indigenous to North America, particularly Virginia, where it occurs very plentifully in damp boggy places. It was introduced into British gardens in 1750.

It is here a compact bush, with numerous short wiry branches, well clothed with lanceolate leaves of a peculiar yellowish-green tint. The flowers are bright yellow, produced in March before the leaves expand, and though not very conspicuous individually, have a pretty effect in the mass, the more valued from the early period of the season at which they appear.

Though an old-fashioned plant, and now rarely met with in collections, it is sufficiently ornamental to be worthy of a place in the American garden; and being quite hardy and of free growth in the ordinary peat-beds, should never be altogether absent in even a moderate collection. It prefers a moist shady situation, and a deep peaty soil.

ERICA (THE HEATH).

Compared with the many hundreds of species and varieties of this brilliant genus, for which we are indebted to the Cape of Good Hope, and which all require greenhouse culture in this country, the European sorts, well known as hardy Heaths, occupy but an insignificant position. They form, nevertheless, a surpassingly beautiful and interesting group of dwarf, free-flowering, evergreen shrubs, easily managed, and worthy of far more attention than has hitherto been bestowed upon them. Of the few species from which the now numerous varieties in cultivation have sprung, the mountains and moorlands of our own country have contributed some of the finest, and they are found in more or less abundance in almost every country in Europe. Growing with the greatest luxuriance in sandy peat, which for the most part forms their natural soil, there are, at the same time, few loams in which they will not succeed, if rich in vegetable matter and free from chalk or lime; while the worst for the purpose may be adapted for their wants by the application of a moderate quantity of peat or old leaf-soil, and even a liberal allowance of well-rotted manure, which they all appreciate very much.

Several of the showiest sorts—such as the varieties of *herbacea*, *mediterranea*, and *australis*—

which flower in the order indicated from February till April, are valuable for winter or spring gardening, and have recently been used with the most admirable results, their neat habit of growth, fresh green foliage, and profusion of bright-coloured flowers, giving a gaiety and effect which no other plants could at that season, and contrasting admirably with the early bulbs with which they are associated.

The other sorts — varieties of *tetralix*, *cinerea*, and *vulgaris*—are in perfection from May to September, the one succeeding the other, when *vagans* begins to develop itself, and continues till late in autumn.

The smaller-growing sorts make neat edgings to beds or borders, as they may be kept trimmed and neat without disparagement to their flowering. The best way, however, of exhibiting their beauty to its fullest extent is that of grouping them in beds by themselves; and when carefully arranged, according to habit and colour of flowers, nothing can be more attractive. To keep them in health and vigour, it is necessary that they should be lifted every four or five years, and either replaced with young plants, which are easily obtained from layers, or sinking the old plants deep enough to cover the bare stems, which render them so unsightly. This can be done with perfect safety, as the young shoots root freely in a few months immediately below the surface. The operation of transplanting may be safely performed at any time between September and April; we prefer, however, the months of February and March, if the weather is open and the ground in good working order, as on the whole the best for the purpose. Most of the sorts are then at rest, and the work is completed before growth commences.

Of a large list of species and varieties, the following are the finest and most distinct :—

E. australis.—This pretty species is found wild in Spain and Portugal, and was first introduced to this country in 1769. It is a close bushy shrub of about 4 feet high, producing its purplish-red flowers in abundance during the summer months. It thrives best in a sheltered situation.

E. ciliaris, indigenous to Portugal, the south of England, and some parts of Ireland, is a neat dwarf species from 9 to 12 inches in height. It produces its pale-red flowers in terminal racemes from June to July. It is one of the finest of the hardy Heaths.

E. cineria.—This species is found in great abundance in many of the northern countries of Europe, and all over Britain, rarely rising above a foot from the ground. The flowers are reddish purple, changing to blue, and begin to expand early in June.

Var. *alba*, occasionally found in the natural

habitats of the species, from which it only differs in having pure white flowers.

Var. *atropurpurea*.—In this variety the flowers are deep rosy purple.

Var. *bicolor*.—In this variety the flowers are pale purple, darker at the point.

Var. *coccinea*, flowers bright red.

Var. *monstrosa*, a curious variety with monstrous flowers.

Var. *pallida*, flowers blush.

Var. *purpurea*, flowers light purple.

Var. *rosea*, flowers bright rose, a very distinct and beautiful variety.

Var. *spicata*, a light pink variety, the flowers produced in spikes.

E. herbacea, indigenous to a wide area in Central Europe, and in some localities in North Wales, is one of the finest of our hardy Heaths. It produces its lovely pale-red blossoms from the beginning of March, and in some seasons much earlier, till the beginning of April; it is a magnificent spring bedding-plant, and as it may be clipped freely without damage, it is valuable as an edging in flower-gardens. It grows about a foot high.

Var. *carnea*.—According to some botanists, this is the type of the species, from which it only differs in having bright red or flesh-coloured flowers; it is also a fine bedding or edging plant.

E. mediterranea, so named from being found abundantly in the countries bordering the Mediterranean Sea. It is also found in several districts in Ireland. In habit of growth and general appearance it resembles the preceding, which, as has been suggested, is probably only a form of this species. Its flowers are pale red, the anthers of a darker colour, and very prominent, and usually in perfection in April.

Var. *alba.*—The flowers of this fine variety are pure white.

Var. *carnea*, flowers flesh-colour.

Var. *glauca.*—In this variety the foliage is glaucous, the flowers similar in colour to the species.

Var. *nana*, a very dwarf form, of a neat, round, bushy habit, well adapted for a small bed, or for edging in miniature flower-gardens.

Var. *stricta*, a very distinct variety, with a close upright habit of growth.

Var. *rubra.*—In this variety the flowers are of a deeper red than those of the species.

E. Mackiana.—This species is indigenous to the Continent, and is also found in Connemara, Ireland. It has broad ovate leaves, silvery on the under surface, possibly a variety of *tetralix;* it grows about a foot high. The flowers are pale red, expanding in July and August. It is a remarkably showy plant.

E. multiflora, indigenous to the south of France, where it grows to a height of about 2 feet. It is a very distinct and showy species, producing its pale-red flowers in great abundance, generally from August to September. It grows very freely, forming a neat bush of about 2 feet high, but requires a sheltered situation.

E. stricta, a native of the mountains of Italy, is a very distinct and handsome species, growing to heights of from 3 to 6 feet. In this country it is rarely found so high, nor does it flower so freely as some of the other species and varieties. Its foliage, for which alone it is worth cultivating, is of a beautiful warm green tint, very delicate and abundant. The flowers are purplish red, and produced in terminal clusters. It frequently suffers damage from our spring frosts, but always recovers in summer.

Var. *minima,* a very dwarf variety, and in point of foliage one of the most elegant of the Heath tribe.

E. tetralix.—This beautiful species is found wild in all the northern countries of Europe, and very abundantly on the moors and heaths of Britain, growing to heights of from 1 to 2 feet. It is readily distinguished by its ciliated leaves, arranged in four whorls round the stems. The flowers are in terminal racemes, of a delicate pink colour, and generally in perfection from July to August.

Var. *rubra.*—In this variety the flowers are of a much deeper red than those of the species.

Var. *alba*, a beautiful variety with white flowers, very showy and desirable.

E. vagans.—This species is found wild in the south of France, in some parts of Ireland, and very abundantly on the moorlands of Cornwall. It grows from 6 inches to 1 foot high, forming a neat compact little bush. The flowers are pale purplish-red, produced in great abundance along the branches. They are generally in perfection in August and September. It is an exceedingly showy plant, invaluable for bedding and margins to clumps of the larger peat-soil shrubs, and forms a neat close edging for flower-gardens.

Var. *alba* has pure white flowers, and is somewhat dwarfer than the species.

Var. *alba nana*, a very dwarf form of the preceding.

Var. *carnea.*—The flowers of this variety are of a deep red or flesh colour.

Var. *rubra*, a very distinct variety with bright pink flowers.

E. vulgaris (*Calluna vulgaris*).—This is the common Heather or Ling of our moors, a plant so well known as to need no description. It is only noticed here as an introduction to an enumeration of its numerous varieties — most of which are so very beautiful that they should never be overloked in forming a collection of hardy Heaths. They are all sports from the species, and have been found from time to time either associated with it in a wild state or in cultivation in gardens.

Var. *alba* has white flowers, which it produces very abundantly.

Var. *Alportii*, a strong-growing variety, producing large spikes of deep red flowers. It is a singularly beautiful plant.

Var. *aurea*, or *ignea*, a very distinct variety with golden variegated foliage.

Var. *argentea*.—In this variety the leaves and branches are freely variegated with a silvery tint.

Var. *coccinea*.—Flowers very dark red.

Var. *decumbens*.— So named from its peculiarly spreading, prostrate habit of growth. Flowers purplish red.

Var. *dumosa*.—This variety has a very dense, bushy habit of growth. Flowers light purplish-red.

Var. *flore pleno* has a robust habit of growth, and double flowers of a pale purplish-red colour. It is a fine showy plant.

Var. *pygmea*, a curious little plant, forming a neat, round, cushion-like bush, with short rigid branches, and small flowers, usually produced very sparingly.

Var. *pumila*, of a dwarf compact habit, but otherwise like the species.

Var. *Hammondii*, a robust variety, bearing long spikes of snowy-white flowers : it is one of the most beautiful of the tribe, and ought to be very extensively cultivated.

Var. *rigida*.—Branches short and rigid; flowers pale purplish-red ; a very distinct variety.

Var. *Searleii*, a neat compact plant with white flowers, which it produces in great abundance ; one of the best of the hardy Heaths.

Var. *tomentosa*.—This form occurs in great abundance in the southern counties of England, and differs from the species in its leaves and branches being covered with a minute down. The flowers are light purplish-red, and produced very abundantly in long spikes.

EMPETRUM (THE CROWBERRY).

Of this family of diminutive evergreens, little more need be said than that, though wanting in that showy appearance which renders some of their allies such universal favourites, they are all pretty little plants, and interesting as varieties in a collection, forming useful edging or permanent bedding plants, their dense procumbent habits of growth eminently fitting them for such purposes. In foliage and general appearance they resemble the Heaths, with which they were formerly associated. They all bear edible berries in considerable abundance, which are ripe in November. All the kinds grow freely in the ordinary peat-beds, but prefer a damp shady situation. Among the species and varieties generally grown we select the following two as being on the whole the most desirable.

E. nigrum, the well-known Crowberry or Crakeberry of our moors, the badge of the clan Maclean, is also found distributed over a wide area in Northern Europe and North America, growing in similar situations. Its small heath-like leaves are dark green, the flowers are of a purplish-white colour, and the berries jet black and very ornamental.

E. rubrum.—A native of the southern coast of South America, from whence it was introduced in 1833, resembles the other species in habit. The leaves and branches have white woolly margins; the flowers are of a dull purple colour, and the berries bright red. This is the finest of the genus, and has a most beautiful effect when associated with other foliage-plants in spring or permanent bedding; it is equally hardy with the European species.

EPIGÆA REPENS (THE MAY-FLOWER).

This, the only species of the genus in cultivation, is a beautiful little creeping evergreen shrub, never rising above the surface of the ground, producing its long, tubular, white, and sometimes light pink, fragrant flowers in great profusion, from May to July. The leaves are of an ovate-cordate shape, about the size of the common Laurestinus, and are, along with the stems, thickly covered with minute hairs.

Though a native of pine-woods, shady rocks, and stony hills in many districts of Canada and the United States, it has been found barely equal to our climate; and unless the situation is peculiarly favourable, requires some such protection as is afforded by a hand-glass or spruce branches during the severer portion of the winter — an indulgence which it richly deserves, as there are few plants more pleasing when in flower, and more deserving of careful attention. It requires a good supply of rough sandy peat, and to be planted in a dry, somewhat shaded situation, such as the east aspect of a rockery, or a bank partially shaded with trees. A few rough pieces of sand or small stones scattered on the surface will be found beneficial, by preventing evaporation and keeping the soil cool, as the Epigæa is very impatient of excessive drought in summer, its roots being extremely delicate, and never penetrating deep below the surface.

Var. *rubicunda*.—This differs only from the species, with which it was found growing in its native habitats, in its flowers being bright red. It is seldom met with now in our gardens, and indeed is probably no longer in cultivation.

GAULTHERIA (THE GAULTHERIA).

A genus of pretty dwarf evergreens, all of them interesting, and, we are convinced, not nearly so widely appreciated by horticulturists as they deserve. The handsome shining foliage and elegant flowers, followed by showy fruit, of several of the species, ought to have secured them far more extended cultivation. Among the various sorts at present known, we select three as being the best and most distinct, and, at the same time, thoroughly hardy in almost any situation :—

G. procumbens (*the Prostrate Gaultheria*), a native of North America from Canada to Virginia, growing in dry woods, on mountains, and on sandy plains. It is a small creeping shrub, with obovate leaves, and white flowers which are produced in July and August. This plant is invaluable for covering the surface of the ground on dry banks or borders shaded with trees ; in such a situation, if liberally supplied with peat, it is most effective, forming a dense carpet of bright foliage, and rarely higher than 6 inches. The fruit, which is a bright scarlet berry, has a sweetish peculiar flavour, is much relished in America, and forms the favourite food of partridges, deer, and many other animals; while the leaves, when properly dried, are used as a substitute for tea.

G. shallon (*the Wood-Laurel*), also from North America, but found growing in damper situations, and often so much shaded that it forms the only undergrowth. It has a procumbent habit, though much more robust than the last species, and produces a dense mass of foliage in woods or shady borders. The flowers, of a pure white colour, are in perfection in May. Apart from its value as an ornamental plant, this species might be introduced extensively with great advantage in woods and shrubberies, for the shelter and food it supplies to game. The berries, of a reddish-purple colour, are produced, where it thrives, in great abundance, and are most delicious as well as wholesome.

G. acuminata (*the Long-leaved Gaultheria*) is a South American species of great beauty, and hardy enough for most situations. The leaves are larger than the last-named sort. It produces its lovely white flowers in May. Though not very often met with in gardens, it is a very ornamental and desirable plant. Like the other species, it prefers a shady situation, but adapts itself to the ordinary beds in the American garden.

KALMIA (THE AMERICAN LAUREL).

Of this genus of American shrubs it is unnecessary to say more, by way of commendation, than that it contains among the few species of which it is composed some of the most handsome evergreens in cultivation, and that they have long been extensively grown and universal favourites both in general shrubberies and in collections of peat-soil plants.

Introduced at intervals between 1734 and 1825 from North America, where they are diffused over a very wide area, growing in rocky woods and high mountain-bogs, some of the species pervading large tracts in a similar manner with the *Calluna vulgaris* of our moorlands, they have proved themselves perfectly hardy and easily cultivated in this country, growing with the greatest luxuriance under the ordinary conditions necessary for the other American plants.

All the species have bright showy flowers, elegant foliage, and neat habits of growth, forming close dwarf bushes, in some cases only a few inches, and never above 4 or 5 feet, in height. The blooming season extends over a considerable period of the spring and summer, some of the sorts coming out early in April, in favourable seasons in March, while others are not in perfection till July.

Along with their attractive colours, the flowers are interesting from their singular shape, the corolla consisting of a single tubular-based and spreading-topped petal, the ten stamens bending back from the centre so as to resemble the ribs of an umbrella, concealing the anthers in an equal number of cavities regularly disposed round the inside; these cavities form horn-like protuberances on the outside, and give to the flower the appearance of an elaborately-carved and elegant salver.

While all the species are really evergreen, some of them have a tendency, in very severe winters or in exposed situations, to partially shed their leaves in mid-winter or early in spring, before the growth begins; this, however, has no bad effect on the health of the plants, as on the approach of genial weather they begin to grow with their usual vigour, and expand their blossoms at the proper season.

Though light sandy peat is the best soil for all the sorts, they will thrive in a rich arenaceous loam, provided it contains plenty of fibre.

The Kalmias are admirable subjects for pot-culture, as they can be easily forced into early flowering, and are most effective and pleasing additions to the decoration of the conservatory at any season. For this purpose they may be potted, and placed at once in the forcing-house. If wanted very early, however, the best plan is to lift them in autumn, or as soon as it is possible to determine which of the plants are best set with flower-buds, and to protect them from frost till it is time to introduce them into heat.

The following species and varieties are distinct and desirable:—

K. angustifolia (*the Narrow-leaved Kalmia*).—This pretty species was introduced so long ago as 1736 from North America, where, particularly in Carolina and Pennsylvania, it is found in great abundance, inhabiting bogs, swamps, and occasionally high mountain-lands. It rarely exceeds 3 feet in height, forming a thickly-branched bush, with myrtle-like leaves of a peculiar light shiny-green colour, in themselves very ornamental, and contrasting well with the darker greens of most other shrubs. The flowers, which are in perfection about the end of June or beginning of July, are of a delicate pink colour, and are produced in bunches from the sides of the branches. It is perfectly hardy, and will thrive in any situation along with the other peat-soil shrubs. Of several well-marked varieties in cultivation the following are the finest:—

Var. *rubra* differs from the species in its flowers being of a darker red colour.

Var. *pumila.*—This is a neat-growing miniature form, but is otherwise similar to the species.

K. glauca (*the Glaucous-leaved Kalmia*), indigenous to Canada and some parts of the United States, from whence it was introduced in 1767, is a beautiful and well-known species, growing from 1 to 2 feet high, densely bushy in its habit, and with dark-green leaves, silvery on the under side; they are much smaller than those of the preceding species. The flowers are pale red or pink, and produced in terminal corymbs. It is a most profuse bloomer, and is generally in perfection in the beginning of April, though in mild springs it is sometimes seen in full flower in the middle of March. It is the best of the family for very early forcing, and may be had in full flower early in January with a moderate amount of heat. A moist situation should always be chosen for this species; and where such cannot be had naturally, it will be found advantageous to supply it liberally with water once or twice during the dry season.

Of varieties, the following are very fine, and ought to find a place in every American garden:—

Var. *stricta*, a well-marked, upright-branched form, otherwise very like the parent.

Var. *superba.*—This variety has much larger and higher coloured flowers, and a slightly more robust habit of growth, than the species.

K. latifolia (*the Calico Bush or Mountain Laurel*), indigenous to Canada and a large area of the colder parts of the United States, inhabiting dry rocky places, and growing to the height of about 8 feet. It was first introduced into this country in 1734, and has been found perfectly hardy in almost every district, forming a close symmetrical bush, abundantly clothed with very dark, glossy green, ovate-lanceolate leaves, resembling both in form and size those of the Sweet-bay. The flowers are much larger than those of any of the other species, of a bright pink colour when newly expanded, but changing afterwards to a paler tint. They are produced in large bunches at the ends of the branches, and are usually in perfection about the end of June.

It is one of the most ornamental of our peat-soil evergreen shrubs, and by far the showiest of the genus to which it belongs.

The situation best adapted for its growth is one fully exposed to the sun. It is impatient of excessive moisture, and care should be taken to have the ground sufficiently drained to prevent the possibility of water stagnating at the roots, as under such a condition it soon assumes a sickly stunted appearance, and rarely, if ever, forms flower-buds. Where the subsoil is cold and damp, the bed in which it is proposed to plant it should be raised above the surrounding surface with sandy peat. When making its growth, however, root-moisture is absolutely necessary; and in naturally dry situations, an occasional copious watering will be found beneficial.

The following varieties are no less beautiful than the species, and thrive well under similar circumstances:—

Var. *major.*—This is a robust form, with larger flowers, but otherwise very similar to the parent.

Var. *myrtifolia*, a charming miniature variety, with a much dwarfer habit of growth, and with foliage and flowers considerably smaller than the species. It is a neat little bush, very useful for front rows of beds or clumps of the more robust American plants, and is now a great favourite with cultivators.

LEDUM (WILD ROSEMARY).

The species which compose this genus are low-growing shrubs, found wild in the northern parts of Europe and North America, inhabiting the most exposed situations, and frequently at high elevations. All the sorts in cultivation have been found to stand the most severe winters in this country without the slightest injury, and to succeed well when planted in a shady, dampish situation, with the usual allowance of turfy peat, or in the case of the natural soil being fibry loam, the addition of a moderate allowance of vegetable mould.

Though less showy in foliage, and to some extent lacking in that brilliancy of colour of flowers which characterises many of the other peat-soil shrubs, the *Ledums* are nevertheless a most ornamental and interesting family of hardy evergreens. Blended as they usually are with the allied plants of a similar height and habit of growth in the American garden, their distinct appearance gives a character and variety to the mixed border or clump which is at once effective and pleasing, and which no one who has so seen them would willingly dispense with.

They are all free bloomers, and, in congenial circumstances, produce their pretty white blossoms in April or May, in great profusion.

L. buxifolium (*the Box-leaved Ledum*), known also as *Ammyrsine buxifolia* and *Leiophyllum thymifolium*, but most commonly as *Ledum*, is a small shrub of about a foot high, indigenous to various parts of North America, particularly on high mountains in Carolina, from whence it was first sent home in 1736. It has a neat, compact habit of growth, the branches very numerous, and thickly clothed with small, ovate, acute, shining-green leaves. The flowers are produced very profusely in terminal corymbs, of a pretty bright pink colour while in bud, and changing when fully expanded to pure white. This, with its varieties, form fine rockwork plants, and has an admirable effect in beds associated with the Heaths and other shrubs of similar habits of growth. They are excellent edging plants to borders or clumps in the American garden, seldom requiring trimming, and growing as neat and dense as the common dwarf Box. They are found naturally in drier and more elevated situations than the other sorts, and in cultivation are found to succeed best in light sandy peat, with the ground well drained.

Var. *thymifolium.*—This is a miniature of the species, differing only in its dwarfer, more compact habit of growth, and smaller foliage.

Var. *thymifolium variegatum.*—This form differs from the last only in having its leaves margined with a yellow variegation.

Var. *intermedium* resembles the parent, but is more robust in habit, and sufficiently distinct and interesting for a mixed collection.

L. latifolium (*the Broad-leaved Ledum*) is a small shrub of about 4 feet in height, indigenous to swamps in Greenland, Canada, and over an immense area of the coldest regions of America. It has been cultivated in Britain since 1763. The leaves are linear-oblong, dark green above, and clothed on the under surface with minute brown tomentum. They have an aromatic fragrance when bruised, and being regarded as pectoral and tonic, as well as agreeably tasted, are sometimes used as a substitute for tea, and the plant is popularly known in some districts of America as *Labrador Tea*. The flowers, which are in terminal corymbs, are pure white, and generally expand in April. Planted in a damp, shady situation, with plenty of bog-peat, it is a remarkably interesting and ornamental shrub, growing freely, and quite hardy enough for our coldest winters.

L. palustre (*the Marsh Ledum or Wild Rosemary*), found wild in Denmark, Silesia, and other countries of Northern Europe, and abundantly over a large portion of Canada and the United States, inhabiting swamps, sides of rivers, and margins of lakes, forming a neat dwarf bush of about 2 feet high. It was introduced in 1762.

The leaves are linear-oblong, with revolute margins, dark green above, and thickly covered on the under side with brown tomentum; when bruised they emit a strong but pleasant aromatic odour. The flowers are pure white, produced in terminal corymbs, and expand generally in May. It is a neat little shrub, similar in style of growth to *latifolium*, but quite distinct enough to be associated with it in collections. It is perfectly hardy, and of easy culture if planted in damp peaty ground, seldom failing to flower freely year after year.

Var. *decumbens*.—This is a very dwarf variety, ound naturally in swamps at Hudson Bay and neighbouring regions. It differs chiefly from the species in having a low trailing habit of growth, and in its much smaller foliage. It is a peculiarly interesting, and, when in bloom, a beautiful little shrub, well suited as a companion to the Heaths and other dwarf peat-soil plants. It seems to grow best in a dry, rather than a moist, situation, and has a fine effect on and about rockeries.

MENZIESIA (THE MENZIESIA).

Resembling their near allies, the Heaths, in their neat bushy habits of growth, elegant foliage, and pretty bell-shaped flowers, the tiny shrubs which compose this genus have long been favourites in the American garden. They are found widely distributed over the colder parts of North America, Asia, and Europe, including Scotland and Ireland, inhabiting dry heathy moors, for the most part at high elevations. All the species and varieties in cultivation are very hardy, and grow freely in the ordinary peat-beds, and in loam rich in fibre or other vegetable matter. They are often, and very appropriately, associated with the Heaths, both in beds of the dwarf peat-soil shrubs, and as margins to clumps of those of more robust growth.

The following are among the most distinct and ornamental of the known sorts :—

M. cærulea (*the Scottish Menziesia*), also known as *Phyllodoce taxifolia*, is found wild in some parts of North America, several countries in northern Europe, particularly in Norway, where it is said to be more abundant than the common Heath. It is also found, but very sparingly, on the eastern brow of the mountain called the Sow of Atholl, in Perthshire. It is a very dwarf plant, rarely found higher than from 4 to 6 inches—forming a broad, tufted, evergreen bush, with numerous wiry stems. The leaves are very small, linear-shaped, toothed on the margins, and glossy green. The flowers, which it produces very copiously about the end of June or early in July, are borne on short stalks from the points of the branches. They are large, and of a beautiful purplish-red colour. It is a charming little rockery plant, thriving well in moderately dry positions if well supplied with sandy peat.

M. empetriformis (*the Empetrum-like Menziesia*), indigenous to high peaty moors over a wide area in North America, from whence it was first introduced in 1810, is another dwarf evergreen, also referred by some botanists to the genus *Phyllodoce*. It has a general appearance similar to that of the preceding species, growing as a trailing heath-like bush of about 6 inches high, the shoots thickly clothed with small linear leaves, toothed at the margins, and of a bright shiny green. The pretty pink flowers, which it generally produces in great profusion, expand about the beginning of July. It is a fine rockery plant, and richly deserving of extensive cultivation.

M. globularis (*the Globe-flowered Menziesia*).—This is a deciduous species indigenous to mountains in North America, and found particularly abundant in Virginia and Carolina, from whence it was introduced in 1806. It is here a small shrub, rarely seen higher than about 2 feet, although described as growing in its native habitats to the height of 5 feet. Its habit of growth is broad and compact, the branches very numerous, thickly clothed with large lanceolate leaves of a light-green tint. The flowers, which generally expand about the end of May, are of a light-brown colour. Though this species cannot be regarded as the showiest of the genus, it is nevertheless very interesting, and worthy of more attention than it has hitherto received. It is now rarely met with in collections.

M. polifolia (*St Dabeoc's* or *Irish Heath*), by some botanists called *Dabeocia polifolia*, is a dwarf, evergreen, heath-like shrub, indigenous to mountainous heaths in various districts of Ireland, and particularly abundant on the low granitic hills to the westward of Galway, forming a thick bush of from 1 to 2 feet high. The leaves are ovate, dark shiny green on the upper surface, and downy beneath. The flowers are of a reddish-purple colour, and usually at their best about the end of July. They are produced in terminal racemes. Grown in sandy peat, and in a moderately dry situation, it blooms profusely year after year, and yields in beauty to none of the other miniature peat-soil shrubs. It is a fine rockery or marginal plant, and as it stands pruning or clipping well, it is often employed as an edging plant in the American garden.

The following varieties are no less ornamental than the parent, and equally worthy of cultivation :—

Var. *alba.*—The flowers of this sort are pure white, thick and waxy in substance, produced very copiously, and remain a long time in perfection. It is a singularly showy and desirable little plant.

Var. *atropurpurea.*—In this variety the flowers are of a much deeper shade of purple than those of the species. They are very effective, from their peculiar colour, as a contrast to the other sorts.

Var. *stricta* or *globosa.*—This is a white-flowering variety, differing from the others in having a more compactly globular habit of growth. It is a neat little bush, and should always be included in the formation of a collection of American plants.

Var. *nana.*—A very dwarf form, free flowering, and when in bloom exceedingly pretty. It is a useful plant for a rockery.

MYRICA (THE CANDLEBERRY MYRTLE).

With inconspicuous flowers produced in brown scaly catkins resembling those of the Birch, the Myricas would scarcely be admitted to the American garden if floral display were the only passport; they are nevertheless graceful shrubs, with elegant foliage possessing a delightful aromatic fragrance, a quality which of itself should insure them a larger share of attention than they have hitherto received. They grow freely in almost every situation if supplied with a moderate quantity of peat, and have a pretty effect in mixed collections of shrubs, either in the American garden proper or in the ordinary pleasure-grounds. The young twigs blend most beautifully among cut flowers in bouquets, their value for this purpose being enhanced by the scent, which is pleasant to most people. Among the species and varieties known in our gardens, the following may be recommended as the most ornamental:—

M. cerifera (*the Wax-bearing Candleberry Myrtle*) is an evergreen shrub growing from 5 to 8 feet high; a native of swamps in Canada and the United States, having shining green leaves of a lanceolate form, and flowering in this country about the beginning of May. The berries or drupes are ripe in October, and are covered with a pure white waxy matter, which is separated by boiling in water, and used in some parts of America, either alone or mixed with animal fat, for making candles. It is a very ornamental shrub; and it is somewhat remarkable that, though introduced into this country so long ago as 1699, it is now rarely met with in our shrubberies or American gardens. Though hardy, it requires to be planted in a moderately sheltered situation.

Var. *latifolia* is a very handsome plant, with much larger leaves, and a more robust habit of growth, than the species, which it otherwise resembles.

M. californica (*the Californian Candleberry Myrtle*), found wild in California and other regions on the north-west coast of America, and first introduced in 1844, is a handsome bushy evergreen, growing from 8 to 12 feet high. The leaves are fragrant, of a narrow lanceolate shape, and thickly disposed on the branches. The flowers, which are of a light-green colour, come out in July in short axillary spikes, succeeded by small berry-like fruit of a dull red colour, ripe in September.

It is quite hardy in this country, and though, like the other species, thriving best in peat, it grows tolerably well in any rich loamy soil. In the American garden or ordinary shrubbery it is a distinct and handsome plant, though as yet not so often met with as we are convinced it deserves.

M. gale (*the Sweet-Gale* or *Bog-Myrtle*) is a decidnous species found wild over a wide area in North America, Northern Europe, and very abundant on the moors and mountain-bogs of the Scottish Highlands, forming a densely-branched bush of about 4 feet high. It is the badge of the clan Campbell. The leaves are lanceolate, serrated, of a dull green tint, and delightfully scented. The flowers, which are in perfection in June, are brownish green, and not very showy. Though very common, and comparatively seldom seen in cultivation, it is nevertheless a handsome shrub, and worth attention for its neat fragrant foliage alone. It thrives best in damp situations.

PERNETTYA (THE PERNETTYA).

This is a small group of dwarf, neat-growing evergreen shrubs, natives of South America. They are nearly all very hardy in this country, and grow freely in peat-soil or in rich fibry loams, producing their pure white bell-shaped flowers from May to July in great profusion, succeeded by abundance of showy berries which hang till late in autumn, and even, in favourable circumstances, over the whole winter. In a young state they are useful as pot-plants for conservatory decoration in winter, requiring no trouble further than lifting them from the border when the fruit is formed, or even ripe, and introducing them at once to the house, in which the berries will hang longer and have a brighter appearance than when exposed to the frost.

The following are the most desirable of the species and varieties:—

P. angustifolia (*the Narrow-leaved Pernettya*), indigenous to Chili, and first introduced to British gardens in 1840, is a handsome, very densely branched shrub of about 3 feet high. The leaves are much narrower than those of the other species. The flowers are white, and the fruit light pink. It is a valuable, hardy, ornamental evergreen, not over fastidious as to soil and situation, and very effective either in the mixed shrubbery border or in the American garden. It thrives well under the shade of trees, but in such a position does not flower freely.

P. mucronata (*the Spiny-leaved Pernettya*), found wild at Cape Horn, and over a large extent of country in the southern regions of South America, and introduced in 1828, is a bushy spreading shrub of from 2 to 3 feet high.

It is thoroughly hardy here, and of remarkably free growth when planted in rich peaty soil. The leaves are small, thick and leathery in texture, of a sharp-pointed ovate form, and of a deep shiny-green colour. The flowers are generally produced very profusely, followed by an abundant crop of large reddish-purple or pink berries. It is a handsome plant at all seasons, but particularly so in winter, when its brilliant showy fruit is in perfection.

Var. *speciosa.*—This form is of a dwarfer and more compact habit of growth, and has smaller leaves, than the parent. It flowers and fruits very freely.

Var. *speciosa candida.*—This differs only from the preceding in the colour of its berries, which are pure white.

Var. *speciosa major,* a very excellent variety, with a more robust habit and larger fruit than the others. All these varieties are admirably adapted for pot-culture, as they flower freely and show their fruit more conspicuously than the species.

POLYGALA CHAMÆBUXUS (THE MILKWORT).

Of this extensive and popular genus, the only ligneous species hardy enough for open-air culture in Britain is *chamæbuxus,* a pretty dwarf evergreen found wild in several countries of continental Europe, particularly Austria, where in some districts it occurs in great abundance, growing in mountain-forests and on heaths. It was first introduced in 1658. Few of our diminutive peat-soil plants are more attractive and useful, its dwarf spreading habit of growth, never exceeding a few inches in height, elegant, box-like, warm green foliage, and thorough hardiness in the most exposed situations, rendering it valuable as an edging plant in the American garden ; while it may be associated in mixed beds with the Heaths, Menziesias, and other plants of similar habits, with the happiest effects.

With the ordinary soil and treatment necessary for the other American plants it thrives to perfection, and never fails to produce its gay light-yellow flowers profusely during the greater part of the summer. From its trailing habit, it requires, when grown as an edging plant, to be trimmed from time to time to keep it in form. If performed in early spring, immediately before it begins to grow, it pushes out young shoots from the sides of the branches, and soon recovers from even a severe pruning, and blooms copiously during the summer.

RHODORA CANADENSIS (THE CANADIAN RHODORA).

This, the only species of the genus in cultivation, by some botanists called *Rhododendron rhodora*, is a spare-growing deciduous shrub of about 3 feet high, found in a wild state over a large portion of Canada and of the United States, inhabiting peaty bogs and the damp margins of lakes and rivers. The leaves are of an oval shape, dark green above and glaucous beneath. The flowers, which are produced in terminal clusters, are of a light rosy-purple colour, and generally expand early in April—before the appearance of the young leaves. Though introduced into this country more than a century ago, and a well-known and popular plant with the gardeners of the last generation, it does not seem now to receive that attention which, as one of the hardiest and earliest of our flowering shrubs, it deserves. This may probably be accounted for by the fact, that in our ordinary peat-beds it seldom assumes that compact vigorous appearance so much valued in ornamental shrubs. It is nevertheless a very interesting plant, and under favourable circumstances produces its showy blossoms in early spring, year after year, in great profusion.

A damp shady situation, with abundance of peat, are indispensable to its wellbeing. It will, indeed, grow and thrive with an amount of moisture that would prove fatal to the great majority of the other American plants; and when cultivated in dry porous soil, it will be necessary to give it a few good waterings during the growing season. It is one of our best forcing plants, easily flowered in moderate heat so early as January, and if afterwards kept cool and shaded, remains a long time in perfection.

VACCINIUM (THE WHORTLEBERRY).

This is an extensive genus of dwarf evergreen and deciduous shrubs, found in a wild state widely distributed over Europe, Asia, and America, inhabiting damp moors, shady woods, or dry mountain-slopes. Of the species and varieties, a considerable number are hardy in this country, and most of them sufficiently distinct and ornamental to justify their being recommended for admission to the American garden; and while we name a few of those which we regard as most useful, there are many others well worthy of cultivation where space can be afforded for their reception. All the sorts produce, more or less abundantly, wholesome and palatable berries, which are used in a variety of ways in the native localities, and in some cases form important articles of commerce. The usual season of flowering is May or June, the berries being ripe about October. In cultivation, some of the species prefer damp, shady situations—others, such as are dry and exposed; but on the whole, their requirements may be easily supplied in the great majority of grounds set apart for peat-soil shrubs. Several of the dwarfer evergreen species make pretty edging or carpet plants in shady woods and shrubberies; their close habit of growth, shining green foliage, and elegant flowers, rendering them ornamental and attractive at all seasons.

V. arctostaphylos (*the Bear's Grape Whortleberry*), indigenous to mountain-woods near the coast of the Black Sea, is a handsome deciduous species, growing from 6 to 8 feet in height, and first introduced in 1800. The leaves are of an elliptic-acute shape, minutely serrated, bright green above and downy beneath. The flowers are blush white, slightly tinged with purple, and produced in racemes from the wood of the preceding year. The berries are of a fine purple colour. It should always be planted in a damp shady situation.

V. buxifolium (*the Box-leaved Whortleberry*).— This is a neat, dwarf, evergreen species, indigenous to Virginia and other parts of North America, sent home in 1794. The leaves are obovate, deep shiny green, and thick in texture. The flowers are pure white with a faint pink stripe. The berries are bright red, and very showy. It grows well in the ordinary peat-beds if planted in a shady exposure; and as it rarely exceeds 6 inches in height, is a fine plant for an edging.

V. frondosum (*the Leafy Whortleberry*), indigenous to woods in New Jersey and Carolina, where it is popularly called *Blue Tangles*, has been cultivated here since 1761. It is a neat deciduous shrub of about 3 feet high, the branchlets short, very numerous, and thickly clothed with obovate-oblong leaves, about 2½ inches long, bright green above, and glaucous, sprinkled with minute dots, on the under side. The flowers are small, greenish white, and produced in drooping racemes from the shoots of the preceding year. The berries are bright blue. It requires a damp shady situation, and a liberal supply of rich peaty soil.

V. macrocarpum (*the American Cranberry*).—This well-known plant, by some writers called *Oxycoccus macrocarpa*, is found wild over the greater part of Canada and the United States, in almost every situation which supplies peat and moisture: was introduced to our gardens about 1760. It resembles very much its ally, the common Cranberry of our mountain-bogs—being a low, trailing, heath-like evergreen shrub, rising only about 6 inches from the ground. The leaves are very small, elliptic-oblong, bright green above and glaucous beneath. It begins to open its pretty pink flowers in May; and the large reddish-purple berries, though ripe in October, hang on the plant during the greater part of the winter. In the American garden it forms a fine contrast in mixed beds with the different varieties of Heaths and other dwarf shrubs, and is an excellent edging plant when kept trim by periodical pruning. It is also sometimes cultivated for its fruit, which is useful for tarts, preserves, and similar purposes, and which it produces freely, when planted in sandy peat, with a situation either naturally moist, or where it can be thoroughly watered two or three times in the course of the summer.

V. ovatum (*the Ovate-leaved Whortleberry*), indigenous to the north-west coast of America, from whence it was introduced in 1826, is a beautiful evergreen shrub of about 2 feet high, with a dense bushy habit of growth. The leaves are ovate-acute and serrated, thick in texture, and of a bright shiny-green colour. The flowers are pink, produced in racemes, and are very showy when fully expanded in May. The berries are black, about the size of currants, and ripe in September. It is an exceedingly distinct and interesting plant, thriving best in a shady aspect, and, like all the other sorts, succeeds best when supplied with plenty of peat.

V. stamineum (*the Long-stamened Whortleberry*), an interesting deciduous species, introduced in 1772 from the swamps of New England, where it grows to a height of about 2 feet. It is a close, short-branched bush, with large, elliptic-acute leaves, beautifully glaucous, the under side covered with a light-coloured down. The flowers are produced in racemes, nearly pure white, and are very showy. The berries when ripe are very light green, almost white. It grows freely, and forms a neat ornamental shrub in a moist shady situation.

V. vitis-idæa (*the Red Whortle or Cow Berry*).—This

is a dwarf evergreen species, found widely distributed over the colder parts of North America, the northern countries of Europe, and abundantly on most of the mountain-heaths of Britain, forming a close dwarf bush rarely exceeding a few inches in height. The leaves are obovate, resembling those of the common Box; they are thick and leathery in texture, and of a bright glossy-green colour. The flowers are pale red or pink, borne in terminal racemes, and in favourable circumstances produced very abundantly. The berries are dark red, very ornamental, and, like those of many of the other sorts, hang on the plants for months after being ripe.

It is one of the prettiest of our carpet or marginal plants, and forms a strong and effective edging to walks when supplied with peat-soil, growing as freely and standing trimming as well as the dwarf Box. A light sandy or gritty peat suits it best, but it succeeds well in the ordinary peat-beds.

Var. *variegata*, a very pretty variety with a straw-coloured variegation, as easily cultivated as the species, and suitable for similar purposes. It is a choice rockery plant.

Var. *major*.—This is an American form, with larger leaves and a more robust habit of growth than the species, with which it is otherwise identical. It is quite hardy, grows freely, and is very desirable as a variety in borders or beds of the dwarf peat-soil shrubs.

HERBACEOUS AND OTHER FLOWERING AND FOLIAGE PLANTS

SUITABLE FOR ASSOCIATING WITH PEAT-SOIL SHRUBS IN THE AMERICAN GARDEN.

———◆———

ALONG with the Rhododendrons, Azaleas, and other shrubs usually grouped together in the formation of what is popularly termed the "American Garden," there are many other highly ornamental plants which, though cultivated with more or less success in ordinary garden soils, thrive best, or at least quite as well, in such as are either wholly or partially composed of peat, and which are sometimes introduced into such arrangements with great advantage; enhancing and prolonging the floral display, and imparting by their distinct, showy foliage a richness and beauty which never fail to interest and please, and which could not be produced without such aids. If there is a weak point in a garden specially set apart for American shrubs, it is the paucity of bloom in midsummer and autumn; and this fact should be kept in view in making a selection of herbaceous plants intended to supplement its ordinary occupants. The following list of the more prominent of the sorts suitable for this purpose might easily be considerably extended, as there are many others no less interesting both in point of foliage and flowers, which will readily suggest themselves to the cultivator who has a desire to improve his collection, and which will secure for him a succession of bloom from the close of the Rhododendron season in June, with little or no interruption, till the following spring, when the early-flowering shrubs begin to display their beauties.

Alstræmeria.—A genus of tuberous-rooted plants of remarkable beauty, but in some cases too tender for our winters in the open ground; several of the species and varieties, however, are hardy, and even the most delicate will stand in ordinary winters if planted from 6 to 9 inches deep in a sunny border, protected with a slight surfacing of litter or leaves.

A. aurea.—This species is the hardiest of the group, growing about 4 feet high, and producing its lovely orange-coloured blossoms in June and July.

A. chilensis.—This and its varieties are among the showiest of the Alstrœmerias; they grow to heights of from 2 to 3 feet, and produce their flowers of various shades of red and yellow very copiously from June to August.

A. psittacina.—Grows about 4 feet high, and produces its crimson flowers in June, July, and sometimes as late as August. It is a strikingly beautiful plant.

Arundo.—This is a genus of grasses with robust stately habits of growth, and of easy culture in any rich garden soil, but thriving with great luxuriance in peat, forming striking objects when judiciously introduced into borders or beds of American shrubs.

A. donax variegata.—A majestic plant of from 6 to 9 feet high; valued for its elegant leaves, which are often from 1½ to 2 feet long, about 2 inches wide, and beautifully striped or margined with a silvery variegation.

A. conspicua.—This species resembles the Pampas Grass in flowers, foliage, and habit of growth, but is much more slender and elegant, besides blooming earlier in summer.

Arum.—This genus contains a number of very interesting decorative plants, distinct in foliage and flowers; they are all of easy culture, and several are hardy enough for our winters in the open ground.

A. cornutum.—A handsome foliage-plant of from 1 to 2 feet high, with curious-shaped, brownish-coloured flowers expanding in May. It is worth cultivating, however, for its foliage alone, which is attractive during the whole of the summer months.

A. dracunculus.—This species has large, spreading, palm-like leaves; the stem grows about 2 feet high—it is prominently spotted; the flowers expand in July, but being of a dullish-brown colour, are more curious than beautiful.

A. italicum has long lance-shaped leaves, and pale-yellow flowers expanding in June; it rarely grows above a foot high. This species and its varieties are very desirable as contrasts in small beds or rockeries among the dwarf American shrubs.

Asclepias.—Most of the species and varieties of this genus are sufficiently ornamental both in foliage and flowers for admission to the American gardens; they are nearly all quite hardy, and luxuriate in peat-soil.

A. incarnata.—A pretty herbaceous plant of about 2 feet high; the flowers are purple, and generally in perfection in July.

A. tuberosus.—This showy and very desirable species grows from 3 to 4 feet high, and produces its fine orange-coloured blossoms in August.

Bulbocodium vernum.—This is one of the earliest of our spring-flowering bulbs, producing its pretty rosy-purple flowers in January or February, according to the mildness of the season, long before the appearance of the leaves, which only begin to develop as they decay. It resembles the Crocus, and is invaluable for planting in beds of the dwarf peat-soil shrubs.

Campanula.—Of this very extensive genus of showy, free-flowering, herbaceous plants, a large proportion grow well and flower profusely in peaty soil; they are nearly all of free growth, and quite hardy in ordinary situations. They vary in height from a few inches to 4 or 5 feet, and are admirable for midsummer and autumn decoration.

C. cæspitosa.—A singularly pretty little plant of about 6 inches high, forming a neat close tuft, covered with blue bell-flowers expanding in July.

C. grandis.—This grows about 2 feet high, producing its large showy blue flowers from July to September.

C. grandiflora.—A handsome species, growing from 1½ to 2 feet high, and producing its magnificent deep-blue flowers in September. There are several varieties with white, striped, and double flowers, equally valuable for decorative purposes.

C. persicifolia.—This is a fine free-growing species, from 2 to 3 feet high, producing its large blue flowers from July to September. Of several varieties the most desirable are *alba*, white-flowered; *alba plena*, double white; *cærulea*, dark blue; *cærulea plena*, double; *coronata alba*, large-flowered white; *coronata cærulea*, large blue.

C. pumila.—A small species of about 4 inches high, forming a close tuft, and producing its pale-blue flowers very copiously in June and July. The variety with pure white flowers (*pumila alba*) is no less interesting. They are most effective plants for small beds or front rows of borders, and blend well with the peat-soil shrubs usually grown in rockeries.

Colchicum autumnale.—This is allied to and resembles the *Bulbocodium vernum*, and is well known as the *Autumn Crocus*. It is an exceedingly beautiful little plant; flowering in November and December, when most other plants are at rest: the flowers vary in colour from purple to pale lilac and white; the leaves are developed in early summer, and decay before the flowers appear.

Crocus.—All the species and varieties of this genus, with their brilliant yellow, blue, white, and striped flowers, are attractive, and coming out as they do in early spring, deserve to be extensively planted among the dwarf shrubs and on the margins of beds of the taller species.

Gentiana.—This genus comprises some of the most beautiful of our hardy herbaceous plants. We notice, however, only such as prefer peat-soil.

G. gelida grows from 9 to 12 inches high, and produces its showy light-blue flowers in June and July very profusely.

G. pneumonanthe.—This plant is found wild on

several of the English heaths growing from 6 to 9 inches high, and producing its spikes of lovely pale-blue flowers in July and August.

Gladiolus.—The grand species, with their now innumerable varieties, which compose this genus, are so well known as to render description superfluous. They are among the gayest of our autumn-flowering plants, and admirably adapted for decorating the American garden, either as massed by themselves, or mixed in beds or borders with the Rhododendrons and other strong-growing shrubs, where, without interfering with their growth, they have a beautiful effect—continuing from July, when *ramosus* and its varieties begin to bloom, till the last flower-spikes of the *Gandavensis* section are cut down by the autumn frosts.

Gynerium argenteum, popularly called the Pampas Grass, is an elegant and majestic plant, producing its beautiful, plume-like, silvery-white flowers about the end of July—all the more valued from their continuing in perfection till the beginning of winter. As it is a robust-growing plant, it should have a bed devoted to itself, or at least ample space on a large border, where its fine foliage and symmetrical outlines will be fully developed and seen to advantage.

Galanthus.—Interesting not only for their beautiful flowers, but for the early period of the season in which they appear, this family of bulbs is worthy of a place in every garden. They thrive well in ordinary garden soil, and accommodate themselves to the peat-beds with equal facility.

G. nivalis.—This is the common Snowdrop, the "herald of the spring," too well known and appreciated to require either description or recommendation.

G. plicatus.—The Crimean Snowdrop should be far more extensively planted than it is at present. It is one of the prettiest of spring-flowering bulbs in cultivation, growing about 9 inches high, and producing its large snow-white flowers about April.

Helleborus.—Dwarf herbaceous plants with showy flowers; important to cultivators from blooming in winter or very early in spring.

H. abchasica.—A pretty little plant of about 1 foot high, with dark-purple, cup-shaped flowers, expanding in February and March.

H. niger, well known as the Christmas Rose, is a small plant of a few inches high, with showy-white flowers expanding in January or February.

H. niger major.—This form is similar to the species, but more robust in habit of growth, and with much larger flowers and foliage.

Hepatica.—This is a small genus of charming little spring-blooming perennials, thriving well in the peat-beds when planted in a shady situation. They have a beautiful effect in March when covered with their bright-coloured flowers, and should be abundantly grown in every American garden.

H. angulosa.—Though not yet very plentiful, this is a most desirable plant. It is readily distinguished from the other species by its much more robust habit of growth and larger flowers. They are of a bright sky-blue colour, borne on stalks of from 6 to 9 inches high, very profuse when the plants are vigorous, and in ordinary seasons at their best about the beginning of March.

H. triloba.—This and its varieties, with white, blue, red, and double flowers, are invaluable for spring decoration, and too well known to require description.

Leucojum vernum.—A March flowering bulb of about 9 inches high, with large white flowers resembling those of the Snowdrop. It is thoroughly hardy, and well suited for planting in beds of the diminutive shrubs.

Lilium.—Few hardy plants produce more gorgeous flowers, or are more easily cultivated, than the Lilies; and as all the species prefer, or at least grow freely in, a peaty soil, no department of the pleasure-ground is better adapted for their cultivation than the American garden. Though all the known sorts are highly ornamental, and worthy of attention, we content ourselves with noting a few of the more prominent.

L. auratum.—Though usually grown in pots for the decoration of the greenhouse and conservatory, this magnificent species is hardy enough for the open air, and succeeds well when planted about 6 inches deep on dry borders. It grows to heights of from 4 to 5 feet, and has a grand effect in the Rhododendron-beds in July or August. The flowers are very large, cup-shaped, pearly white, the centre of each petal striped with gold, and prettily spotted with chocolate crimson. It is delightfully fragrant.

L. bulbiferum umbellatum, a very beautiful variety of one of the commonest of the species; it varies in height from 2 to $3\frac{1}{2}$ feet, and produces in June, or early in July, splendid umbellate racemes of cup-shaped, orange-red blossoms. Like the species, it is perfectly hardy, and thrives well in almost every variety of soil.

L. candidum.—This well-known and much-admired plant grows to a height of from 4 to 5 feet, producing large racemes of trumpet-shaped, snow-white, fragrant flowers, which are generally in perfection in July or early in August.

L. chalcedonicum—popularly called the Scarlet Martagon Lily—is a very showy species, growing about 4 feet high, and producing its brilliant scarlet turban-like blossoms in July and August. Though a common it is a most desirable plant, well adapted for grouping with shrubs in large clumps or beds.

L. giganteum.—This is a singularly beautiful plant, with a stately habit of growth, rising to heights of from 6 to 8 feet, with large, glossy-green, heart-shaped leaves, and terminal racemes of large, trumpet-shaped, white flowers, streaked with violet crimson, and sweet-scented. They are usually in perfection early in August.

L. longiflorum is a dwarf species, never growing above about 1½ foot high. Its flowers are similar in form to, but much larger than, those of *candidum :* they are snowy-white and delicately scented, and in perfection in August. It is very hardy, of easy culture, and very desirable for planting among the dwarf shrubs.

L. speciosum—sometimes called *lancifolium*—is a splendid species, growing from 3 to 5 feet high, and producing its lovely pink or light-crimson, spotted, turbinate flowers in August and September. There are several equally beautiful varieties, differing from the parent chiefly in the colours and markings of the flowers. Of these the most distinct are *album*, pure white; *punctatum*, white, delicately spotted with pink; *roseum*, bright rose, finely spotted with crimson; *cruentum*, deep rose, heavy crimson spots.

L. testaceum.—A very showy species, with Turkscap flowers of a light-shaded yellow colour, expanding about the beginning of July. It grows from 4 to 5 feet high, and is a magnificent plant for associating with the tall shrubs.

L. Thunbergianum—also known as *venustum*—is a remarkably beautiful species, growing about 2 feet high; the flowers are cup-shaped, of a buff-orange colour, and expand early in August. The varieties—*grandiflorum*, with crimson flowers, and *grandiflorum flore-pleno*, with double crimson flowers—are also gems, and ought to be more frequently met with in collections.

L. tigrinum.—A very common but a very showy plant, growing about 3½ or 4 feet high, and producing its pretty orange, spotted, turbinate flowers in July in great profusion.

L. Washingtonium.—This is a species of recent introduction from Nevada. It grows from 4 to 5 feet high, and produces in August a panicle of large trumpet-shaped flowers, white, shading to lilac. Though as yet almost entirely cultivated in pots, there is little doubt of its being quite hardy, and of its proving a splendid acquisition to our list of border Lilies. Until it is more thoroughly tested, however, it will be prudent to protect it during winter with a slight covering of litter.

Osmunda.—A genus of Ferns of robust growth,

and peculiarly ornamental when planted in damp peaty soil and shady aspects.

O. regalis—popularly called the Royal Fern—is a stately distinct-looking plant, indigenous to peaty bogs in several parts of Britain, growing generally to a height of about 4 feet, but sometimes in peculiarly favourable circumstances found with fronds from 8 to 10 feet long. It thrives well cultivated in the peat-soil beds, and contrasts admirably with the American shrubs. The variety named *cristata* is a beautifully-crested form of the species, and well deserving of cultivation.

O. interrupta.—This is a North American species, differing from the Royal Fern in its dwarfer habit of growth, as well as in its fructification, which, instead of being terminal, as in that species, is produced about the middle of the frond, the barren portions being above and below it. It rarely grows above 3 feet.

O. spectabilis. — Another North American form, resembling *regalis*, but much more slender in all its parts. While young, the fronds are of a purplish tint. It grows about 3 feet high, and is a most graceful and distinct Fern.

Phormium tenax—known as the New Zealand Flax—is a majestic foliage-plant, with an aspect suggestive of a gigantic grass. It thrives admirably in peaty soil, but, except in very mild districts, is scarcely equal to our severest winters without a little protection. There are two or three variegations well worthy of attention as handsome contrasts in the American garden. All the sorts should be planted in a dry but sheltered situation.

Struthiopteris pennsylvanica — known as the "Ostrich Plume" fern—is a magnificent, hardy, North American species, found to grow luxuriantly in damp peaty soil, and shady situations. The fronds in old well-established plants are sometimes found 4 feet high. While young, they are of a beautiful light green, assuming a deeper hue when matured. It is deciduous, but very effective in early spring and summer.

Tritoma uvaria.—This magnificent autumn-flowering plant forms a large dense bush with long grass-like leaves, and throws up flower-stems of from 3 to 6 feet in height, forming at the top a spike from 1 to $1\frac{1}{2}$ foot of closely-arranged, scarlet and yellow, tube-shaped blossoms. It is a very hardy plant, and unrivalled for intermixing with robust shrubs in large clumps, and forms a charming specimen plant on grass. It is usually at its best in October, but is often more or less beautiful till the end of November.

Tigridia.—In this genus there are several species of very showy bulbous plants, of about 1 foot high, flowering in July and August. They are too tender for our winters in the open air, and require to be lifted in autumn and kept in a dry state till April. They do best in a dry sunny border, and produce their beautiful blossoms very profusely.

T. conchiflora has Iris-like foliage, and large yellow flowers, spotted with scarlet.

T. pavonia differs from the preceding in the colour of its flowers—the ground-colour being dark scarlet, and the spots bright orange.

T. speciosa.—In this species the ground-colour is deep crimson and the spots orange.

Trillium grandiflorum.—This is a dwarf perennial of great beauty, producing its attractive white flowers in June and July on stalks about 9 inches high. It will only thrive in damp peaty soils, and in situations shaded from the sun, and is a superb plant for the margins of clumps or beds of American shrubs.

Yucca.—A group of handsome evergreens, with sword-shaped leaves and tall panicles of large tulip-like flowers. They are much valued for their singularly graceful foliage, and give a fine feature to any arrangement of shrubs.

Y. filamentosa.—This is a stemless species, interesting from the curious threads borne on the margins of the leaves. The flowers are white, and in perfection in September.

Y. gloriosa.—A superb species, with noble leaves and a symmetrical Palm-like habit of growth. The flowers are white, and borne on a stem of about 6 feet in height, and when in full perfection have a very imposing effect. It is a fine plant as a single specimen for a prominent position.

Y. recurvifolia.—A remarkably elegant species, with long leaves gracefully drooping at the points. The flowers are white, and usually at their best in September. It is a very ornamental plant for a lawn or vase, and has a fine effect in the centre of a bed of mixed shrubs.

INDEX.

ABIES, 9—Albertiana, 10—alba and varieties, 11—Alcoquiana, 12—Canadensis, 12—Douglasii and varieties, 13 *et seq.*—Englemannii, 17—excelsa and varieties, *ib. et seq.* Menziesii, 20—morinda, *ib.*—nigra, 22—orientalis, *ib.*—obovata, *ib.*—polita, 23—Pattoniana, 24.
African Cypress, the, 167.
Alcock's Spruce, 12.
Alpine Azalea, the, 229.
Alpine Bearberry, the, 224.
Alstræmeria, 274—aurea, *ib.*—chilensis, *ib.*—psittacina, 275.
American Arbor vitæ, the, 150. See Thuja.
American Laurel, the, 248.
American or peat-soil shrubs. general remarks on their culture, 168 *et seq.*
Andes Podocarpus, the, 123.
Andromeda, 217—axillaris, 219—angustifolia, *ib.*—calyculata, *ib.*—floribunda, 220—polifolia, 221—pulverulenta, *ib.*—rosmarinifolia, 222—tetragonia, *ib.*
Araucaria, the, 25—imbricata, 26.
Arbor vitæ, the American, 150—the Broad-leaved, 156 the Eastern, 29.
Arctostaphylos, 223—alpina, 224—uva ursi, *ib.*
Arum, 275—cornutum, *ib.*—dracunculus, 276—italicum, *ib.*
Arundo, 275—donax, *ib.*—conspicua, *ib.*
Asclepias, 276—incarnata, *ib.*—tuberosus, *ib.*
Austrian Pine, the, 83.
Azalea, the, 210—list of varieties, 214.
Azalea (Chamæledon) procumbens, 229.

BALFOUR'S PINE, 85.
Balm of Gilead Fir, the, 111.
Bearberry, the, 223.
Bentham's Pine, 84.
Bhotan Pine, the, 89.

Biota, the, 29—orientalis and varieties, 30—filiformis or pendula, 33—japonica, 34 - meldensis, *ib.*
Bishop's Pine, the, 98.
Black American Spruce, the, 21.
Black Austrian Pine, the, 84.
"Blue John," 145.
Blunt-leaved Retinospora, the, 130.
Bog-Myrtle, the, 262.
Broad-leaved Arbor vitæ, the, 156. See Thujopsis.
Bryanthus erectus, 225.
Bulbocodium vernum, 276.

CALICO BUSH, the, 252.
Californian Nutmeg, the, 160.
Californian Redwood tree, the, 138.
Californian Yew, the, 148.
Calluna vulgaris, 241.
Campanula, 277—cæspitosa, *ib.*—grandis, *ib.*—grandiflora, *ib.*—persicifolia, *ib.*—pumila, 278.
Canadian Yew, the, 147.
Candleberry Myrtle, the, 261.
Cedar, the, 35. See Cedrus.
Cedar, the Japan, 41. See Cryptomeria.
Cedar of Lebanon, the, 35.
Cedrus, 35—libani, *ib.*—deodara, 37 varieties, 39—atlantica, *ib.*
Cephalotaxus, 43—Fortunei, 44—drupacea, 45—pedunculata, 46.
Chamæcyparis sphæroidea, 47—varieties, 48.
Chamæledon procumbens, 229.
Chile Pine, the, 27.
Chilian Libocedrus, the, 78.
Chinese Arbor vitæ, the, 30.
Chinese Juniper, the, 64.
Christmas Rose, the, 280.
Cilician Silver Fir, the, 112.
Clethra, 226—alnifolia, *ib.*—tomentosa, 227.
Clubmoss Retinospora, the, 129.
Cluster-coned Pine, the, 101.
Cluster-fruited Yew, the, 43. See Cephalotaxus.

T

INDEX.

Colchicum autumnale, 278
Common Juniper, the, 61.
Common Larch, the, 73.
Common Silver Fir, the, 118.
Comptonia asplenifolia, 227.
Conifers, Ornamental: Introduction, 1—Abies, 9—Araucaria, 25—Biota, 29—Cedrus, 35—Cryptomeria, 41—Cephalotaxus, 43—Chamæcyparis, 47—Cupressus, 49—Fitz-Roya, 59—Juniperus, 60—Larix, 72—Libocedrus, 78—Pinus, 81—Picea, 107—Podocarpus, 122—Prumnopytis, 125—Retinospora, 126—Salisburia, 134—Saxe-Gothæa, 136—Sciadopytis, 137—Sequoia, 138—Taxus, 140—Taxodium, 148—Thuja, 150—Thujopsis, 156—Torreya, 159—Wellingtonia, 161—Widdringtonia, 167.
Corean Pine, the, 93.
Corean Podocarpus, the, 123.
Corean Spruce, the, 23.
Corsican Pine, the, 95.
Cowberry, the, 271.
Creeping Yew, the, 146.
Crocus, 278.
Crowberry, the, 243.
Cryptomeria, 41—japonica, *ib.*—varieties, 42—elegans, *ib.*
Cupressus, 49—sempervirens, 50—variety, 51—Lawsoniana, *ib.*—varieties, 53—macrocarpa, 54—varieties, 55—Goveniana, 56—M'Nabiana, 57—nutkaensis, *ib.*—varieties, 59.
Cupressus Thyoides (Chamæcyparis sphæroidea), 47.
Cypress, the, 49. *See* Cupressus.

DAPHNE, 230—collina, 231—cneorum, 232.
Deciduous Cypress, the, 148. *See* Taxodium.
Decurrent-leaved Libocedrus, the, 79.
Deodar Cedar, the, 37.
Dirca palustris, 234.
Douglas's Spruce, 13.
Dwarf White Spruce, the, 12.

EASTERN ARBOR VITÆ, the, 29. *See* Biota.
Eastern Spruce, the, 22.
Elegant Cryptomeria, the, 42.
Empetrum, 243—nigrum, 244—rubrum, *ib.*
Engleman's Spruce, 16.
Epigæa repens, 245.
Erica, 235—australis, 237—ciliaris, *ib.*—cinerea, *ib.*—herbacea, 238—mediterranea, 239—Mackiana, *ib.*—multiflora, 240—tetralix, *ib.*—vagans, 241—vulgaris, *ib.*

FERN-LIKE RETINOSPORA, the, 128.
Fitz-Roya Patagonica, 59.
Flat-branchletted Retinospora, the, 129.
Flattened or Creeping Yew, the, 146.
Fœtid Yew, the, 159.
Fortune's Cephalotaxus, 44.
Frankincense Juniper, the, 70.
Fraser's Silver Fir and variety, 113.
Fremont's Nut-pine, 90.
French hybrid Arbor vitæ, 34.

GALANTHUS, 279—nivalis, 280—plicatus, *ib.*
Garland Flower, the, 232.
Gaultheria, 246—procumbens, *ib.*—shallon, 247—acuminata, *ib.*
Gentiana, 278—gelida, *ib.*—pneumonanthe, *ib.*
Gladiolus, 279.
Glaucous-leaved White Spruce, the, 11.
Golden Larch, the, 75.
Golden Yew, the, 145.
Gowen's Cypress, 56.
Great Silver Fir, the, 114.
Ground Cypress, the, 47.
Gynerium argenteum, 279.

HACKMATACH, the, 76.
Hatchet-leaved Thujopsis, the, 156.
Heath, the, 235.
Heath-like Retinospora, the, 127.
Heavy-wooded Pine, the, 100.
Helleborus, 280—abchasica, *ib.*—niger, *ib.*
Hemlock Spruce, the, 12.
Hepatica, 280—angulosa, 281—triloba, *ib.*
Himalayan Spruce, the, 20.

INCENSE CEDAR, the, 78. *See* Libocedrus.
Indian Cedar, the, 37.
Indian Silver Fir, the, 120.
Irish Heath, the, 259.
Irish Juniper, the, 62.
Irish Yew, the, 143.
Italian Daphne, the, 231

JAPAN ARBOR VITÆ, the, 34.
Japan Cypress, the, 126. *See* Retinospora.
Japan Juniper, the, 71.
Jeffrey's Pine, 93.
Juniper, the, 60. *See* Juniperus.
Juniperus, 60—communis, 61—varieties, 62—oxycedrus, *ib.*—virginiana, 63—varieties, *ib.*—drupacea, 64—chinensis, *ib.*—varieties, 65—excelsa and varieties, 66—sabina and varieties, 67—recurva, 68—varieties,

69—squamata, *ib.* —tamariscifolia, 70 thurifera, *ib.* —japonica, 71— rigida, *ib.*
KALMIA,248—angustifolia,250—glauca, 251—latifolia, 252.
Khutrow Spruce, the, 20.
LAMBERT'S PINE, 94.
Larch, the, 72. *See* Larix.
Large-coned Pine, the, 99
Large-fruited Cypress, the, 54.
Larix, 72—europæa, 73—varieties, 74 — Kæmpferi, 75 — microcarpa or Americana, and varieties, 76.
Lawson's Cypress, 51.
Leather-wood, the, 234.
Ledum, the, 253—buxifolium, 254— latifolium, 255—palustre, 256.
Leucojum vernum, 281.
Libocedrus, the, 78—chilensis, *ib.* —decurrens, 79.
Lilium, 281 — auratum, 282 — bulbiferum umbellatum, *ib.* —candidum, *ib.* —chalcedonicum, *ib.* —giganteum, 283—longiflorum, *ib.* —speciosum, *ib.* —testaceum, *ib.* — Thunbergianum, 284—tigrinum, *ib.* —Washingtonium, *ib.*
Long-stalked Cephalotaxus, the, 46.
Long-stalked Yew, the, 122. *See* Podocarpus.
Lovely Silver Fir, the, 110.
Lycopod-like Thujopsis, the, 158.

M'NAB'S CYPRESS, 57.
Maidenhair Tree, the, 134.
Mammoth Tree, the, 161.
Marsh Ledum, the, 256.
Mayflower, the, 245.
Menzies's Spruce, 20.
Menziesia, 257—cærulea, 258—empetriformis, *ib.* —globularis, 259—polifolia, *ib.*
Milkwort, the, 265.
Moorwort, the, 221.
Mount Atlas Cedar, the, 38.
Mount Enos Fir, the, 112.
Mountain Laurel, the, 252.
Mountain Pine, the, 100.
Myrica, 261—cerifera, *ib.* —californica, 262—gale, *ib.*

NEW ZEALAND FLAX, 285.
Noble Silver Fir, the, 116.
Nootka Sound Cypress, 57.
Nordmann's Silver Fir, 117.
Norfolk Island Pine, the, 27.
Norway Spruce, the, 17.
Nubigean Podocarpus, the, 124.
Nut-bearing Torreya, the, 160.

OBOVATE-CONED SPRUCE, the, 22.
Osmunda, 284—regalis, 285—interrupta, *ib.* —spectabilis, *ib.*
Ostrich-plume Fern, 286.

PAMPAS GRASS, 279.
Patagonian Fitz-Roya, the, 59.
Patton's Spruce, 24.
Pea-fruited Retinospora, the, 131.
Peat-soil shrubs, general remarks on culture of, 168—Rhododendrons, 170 — Azalea, 210—Andromeda, 217— Arctostaphylos, 223—Bryanthus, 225 — Clethra, 226—Comptonia, 227— Chamæledon, 229— Daphne, 230— Dirca, 234—Erica, 235—Empetrum, 243—Epigæa, 245—Gaultheria, 246 Kalmia,248—Ledum,253—Menziesia, 257—Myrica, 261—Pernettya, 263— Polygala, 265—Rhodora, 266—Vaccinium, 268. Plants suitable for associating with them, 273—Alstræmeria, 274 — Arundo, 275 — Arum, *ib.* — Asclepias, 276—Bulbocodium, *ib.* — Campanula, 277—Colchicum, 278— Crocus, *ib.* —Gentiana, *ib.* —Gladiolus, 279—Gynerium, *ib.* — Galanthus, *ib.* —Helleborus, 280—Hepatica, *ib.* — Leucojum, 281—Lilium, *ib.* —Osmunda, 284—Phormium, 285 —Struthiopteris, 286—Tritoma, *ib.* — Tigridia, *ib.* —Trillium, 287—Yucca, *ib.*
Pernettya, 263—angustifolia, 264— mucronata, *ib.*
Phormium tenax, 285.
Picea, 107—amabilis, 110—balsamea, 111—cephalonica, 112—cilicia, *ib.* — Fraseri, 113—grandis, 114—lasiocarpa, *ib.* —magnifica, 115—nobilis, 116 — Nordmanniana, 117— pectinata, 118—pichta, 119—pinsapo, 120— pindrow, *ib.* -Webbiana, 121.
Pine, the, 81. *See* Pinus.
Pinetum, the, situation, soil, &c., for it, 2 *et seq.*
Pinsapo Silver Fir, the, 120.
Pinus, 81—austriaca, 83—variety, 84— Benthamiana, *ib.* —Balfouriana, 85— contorta, 86—Cembra, 87—varieties, 88 — excelsa, 89—variety, 90—Fremontiana, *ib.* —insignis,91—Jeffreyii, 93—koriensis, *ib.* — Lambertiana, 94 —laricio, 95—variety, 98—muricata, *ib.* macrocarpa, 99—monticolo, 100 — ponderosa, *ib.* — pinaster, 101— variety, 102—radiata, *ib.* —rigida, 103 —Sabiniana, *ib.* -strobus, 104—sylvestris, 105—varieties, *ib.* —tuberculata, 106.
Pitch Silver Fir, the, 119.

INDEX.

Plaited-leaved Arbor vitæ, the, 154.
Plume-like Retinospora, the, 132.
Plum-fruited Cephalotaxus, the, 45.
Plum-fruited Juniper, the, 64.
Podocarpus, 122 — andina, 123 — koriana, ib.—nubigæna, 124.
Polygala chamæbuxus, 265.
Prickly Juniper, the, 62.
Prince Albert's Spruce, 10.
Prince Albert's Yew, 136.
Prumnopytis elegans, the, 125.

RADIATED-CONE PINE, the, 102.
Red American Larch, the, 76.
Red Cedar, the, 63.
Remarkable Pine, the, 91.
Retinospora, 126—ericoides, 127—filicoides, 128—filifera, ib.—leptoclada, 129—lycopodioides, ib.—obtusa, 130—varieties, 131—pisifera, ib.—varieties, 132—plumosa, ib.—varieties, 133—squarrosa, 133.
Rhododendron, the, 170—hybridising, 174—soils for them, 177—situation, 180—transplanting, 181—manure, 183—propagation and grafting, 184—forcing, 186—selection of kinds, and lists of these, 192.
Rhodora canadensis, 266.
Rose Bay, the, 170. *See* Rhododendrons.
Royal Fern, the, 285.

SABINE'S PINE, 103.
St Daboec's Heath, 259
Salisburia adiantifolia and varieties, 134.
Savin Juniper, the, 67.
Saxe-Gothæa conspicua, 136.
Scaly-leaved Juniper, the, 69.
Scaly-leaved Retinospora, the, 133.
Sciadopytis verticillata, the, 137.
Scots Pine, the, 105.
Sequoia sempervirens, 138.
Siberian Silver Fir, the, 119.
Silver Fir, the, 107. *See* Picea.
Situation for Pinetum, 3.
Small-coned Larch, the, 76.
Snowdrop, the, 279.
Soil for Conifers, 4.
Spanish Juniper, the, 70.
Spruce Firs, the, 9. *See* Abies.
Standish's Thujopsis, 158.
Stiff-leaved Juniper, the, 71.
Stiff-leaved Pine, the, 103.
Stone Pine, the, 87.
Struthiopteris pennsylvanica, 286.
Superb Silver Fir, the, 115.
Swedish Juniper, the, 62.

Sweet Gale, the, 262.
Swiss Stone Pine, the, 87.

TALL ARBOR VITÆ, the, 151.
Tall Juniper, the, 66.
Tamarack, the, 76.
Tamarisk Juniper, the, 70.
Taxodium distichum, 148 — varieties, 150.
Taxus, 140—baccata, 141—varieties, 143—adpressa, 146—canadensis, 147—Lindleyana, 148.
Taxus japonica, 123.
Thread-branched Arbor vitæ, the, 33.
Thread-branched Retinospora, the, 128.
Thuja, 150—gigantea, 151—occidentalis, 152—varieties, 153—plicata, 154—varieties, 155.
Thuja chilensis, 78.
Thujopsis, 156—dolobrata, ib.—varieties, 157—lætevirens, 158 — Standishii, ib.
Thujopsis borealis, 58.
Tigridia, 286—conchiflora, 287—pavonia, ib.—speciosa, ib.
Torreya, 159—nucifera, 160—myristica, ib.
Transplanting Conifers, 5.
Trillium grandiflorum, 287.
Tritoma uvaria, 286.
Tsuga Lindleyana, 14.
Tuberculated-coned Pine, the, 106.
Twisted-branched Pine, the, 86.

UMBRELLA PINE, the, 137.
Upright Cypress, the, 50.
Upright Indian Silver Fir, the, 120.

VACCINIUM, 268—arctostaphylos, 269—buxifolium, ib.—frondosum, ib.—macrocarpum, 270—ovatum, 271—stamineum, ib.—vitis-idæa, ib.

WEBB'S SILVER FIR, 121.
Weeping Juniper, the, 68.
Wellingtonia gigantea, the, 161.
Western Arbor vitæ, the, 152.
Weymouth Pine, the, 104.
White Cedar of America, the, 47.
White Spruce, the, 11.
Whortleberry, the, 268.
Widdringtonia cupressoides, 167.
Wild Rosemary, the, 253.
Wood Laurel, the, 247.
Woolly-scaled Silver Fir, the, 114.

YEW, the, 140. *See* Taxus.
Yucca, 287—filamentosa, ib.—gloriosa, 288—recurvifolia, ib.

WORKS ON GARDENING.

DOMESTIC FLORICULTURE, WINDOW-GARDENING, and FLORAL DECORATIONS: Being Directions for the Propagation, Culture, and Arrangement of Plants and Flowers as Domestic Ornaments. By F. W. BURBIDGE, Author of 'Cool Orchids, and how to Grow them.' In crown 8vo, with 200 Illustrations, 7s. 6d.

"This book will meet the case of thousands who love flowers, and know not how to begin—or, having begun, know not how to go on in collecting and cultivating them. . . . It is a model of painstaking accuracy and good taste."—*Gardeners' Magazine.*

"Such an amount of instruction, information, and illustrations, as has never before appeared in a handbook of this description. . . . We most heartily commend the work."—*The Gardener.*

"This book may be described as a good idea well carried out, and will assuredly commend itself to the amateur class for whose use it is provided."—*The Florist and Pomologist.*

HANDY BOOK OF THE FLOWER-GARDEN: Being Practical Directions for the Propagation, Culture, and Arrangement of Plants in Flower-Gardens all the Year Round. Embracing all classes of Gardens, from the largest to the smallest. With Engraved and Coloured Plans, illustrative of the various Systems of Grouping in Beds and Borders. By DAVID THOMSON. A New and Enlarged Edition, crown 8vo, 7s. 6d.

"To sum up, this 'Handy Book' deserves a welcome from all classes interested in Floriculture."—*Saturday Review.*

"One of the very few books of its kind in which the amateur, instead of being overwhelmed by details, has the principles which are to guide him put plainly and clearly before him, so that he may be able to think and judge for himself."—*Pall Mall Gazette.*

"It is a practical volume, which we recommend to our readers without any reservation."—*Journal of Horticulture.*

HANDBOOK OF HARDY HERBACEOUS AND ALPINE FLOWERS FOR GENERAL GARDEN DECORATION: Containing Descriptions, in Plain Language, of upwards of 1000 Species of Ornamental Hardy Perennial and Alpine Plants, adapted to all Classes of Flower-Gardens, Rockwork, Groves, and Waters; along with Concise and Plain Instructions for their Propagation and Culture. By WILLIAM SUTHERLAND, formerly Manager of the Herbaceous Department at Kew. Crown 8vo, 7s. 6d.

"Mr Sutherland's volume is second to none in its honest work, valuable hints, and compact practical information."—*Saturday Review.*

A PRACTICAL TREATISE ON THE CULTIVATION OF THE GRAPE VINE. By WILLIAM THOMSON, of Tweed Vineyard, Galashiels. Eighth Edition, enlarged. 8vo, 5s.

WORKS ON GARDENING—continued.

A New and Enlarged Edition.
A BOOK ABOUT ROSES: How to Grow and Show them. By the Rev. S. REYNOLDS HOLE, Author of a 'Little Tour in Ireland.' Fifth Edition, 7s. 6d.

"The whole volume teems with encouraging data and statistics; and, while it is intensely practical, it will interest general readers by an unfailing vivacity, which supplies garnish and ornament to the array of facts, and furnishes 'ana' in such rich profusion that one might do worse than lay by many of Mr Hole's good stories for future table-talk."—*Saturday Review.*

"It is the production of a man who boasts of thirty 'all England' cups, whose Roses are always looked for anxiously at flower-shows, who took the lion's share in originating the first Rose-show *pur et simple*, whose assistance as judge or *amicus curiæ* is always courted at such exhibitions. Such a man 'ought to have something to say worth hearing to those who love the Rose,' and he *has* said it." *Gardeners' Chronicle.*

THE SIX OF SPADES: A Book about the Garden and the Gardener. By the Rev. S. REYNOLDS HOLE, Author of 'A Book about Roses,' &c. Crown 8vo, 5s.

"We may, in conclusion, recommend the whole book to the attention of our readers as one which will afford them much amusement on a winter's night.... It is written by one who really loves flowers, and wishes to lead others to worship at the same shrine; and we wish the book success."—*Journal of Horticulture.*

ON ORNAMENTAL-FOLIAGED PELARGONIUMS. With Practical Hints for their Production, Propagation, and Cultivation. By PETER GRIEVE, Culford, Bury St Edmunds. Second Edition, enlarged, including description of Best Varieties introduced up to the present time, and Engravings. Crown 8vo, 4s.

THE HANDY BOOK OF FRUIT CULTURE UNDER GLASS. By DAVID THOMSON, Author of 'Handy Book of the Flower-Garden,' 'A Practical Treatise on the Culture of the Pine-Apple,' &c. In crown 8vo, with Engravings, 7s. 6d.

"We can recommend this work very highly to all who are engaged in fruit culture under glass as a thoroughly practical and reliable guide."—*Journal of Horticulture.*

"The author is well known to be a thorough master of his profession, and one of the most able and best practical gardeners of the present day. We therefore expected, on opening this volume, to find it brimful of good sound practical advice, and we have not been disappointed. The work before us is a true gardener's book."—*Gardeners' Chronicle.*

THE BOOK OF THE GARDEN. By CHARLES M'INTOSH, formerly Curator of the Royal Gardens of his Majesty the King of the Belgians, and lately of those of his Grace the Duke of Buccleuch, K.G., at Dalkeith Palace. In 2 large vols. royal 8vo, embellished with 1350 Engravings. £4, 7s. 6d.

WILLIAM BLACKWOOD & SONS, EDINBURGH AND LONDON.

www.ingramcontent.com/pod-product-compliance
Lightning Source LLC
Chambersburg PA
CBHW030748250426
43672CB00028B/1320